Do I Count?

Stories from Mathematics

Do I Count?
Stories from
Mathematics

Günter M. Ziegler

Translated by Thomas von Foerster

CRC Press
Taylor & Francis Group
Boca Raton London New York

CRC Press is an imprint of the
Taylor & Francis Group, an **informa** business

AN A K PETERS BOOK

Originally published in German as *Darf ich Zahlen?: Geschichten aus der Mathematik* © 2010 by Piper Verlag GmbH, Müchen

CRC Press
Taylor & Francis Group
6000 Broken Sound Parkway NW, Suite 300
Boca Raton, FL 33487-2742

© 2014 by Taylor & Francis Group, LLC
CRC Press is an imprint of Taylor & Francis Group, an Informa business

No claim to original U.S. Government works

Printed on acid-free paper
Version Date: 20130424

International Standard Book Number-13: 978-1-4665-6491-6 (Paperback)

Library of Congress Cataloging-in-Publication Data

Ziegler, Günter M.
  [Darf ich Zahlen? English]
  Do I count? : stories from mathematics / Günter M. Ziegler ; translated by Thomas von Foerster.
    pages cm
  "An A K Peters book."
  "Originally published in German as Darf ich Zahlen? : Geschichten aus der Mathematik ... 2010 by Piper Verlag GmbH, Müchen"--Title page verso.
  Includes bibliographical references and index.
  ISBN 978-1-4665-6491-6 (paperback : acid-free paper)
  1. Mathematics--Anecdotes. 2. Mathematicians--Anecdotes. I. Title.

QA99.Z5413 2013
510--dc23   2013009573

Visit the Taylor & Francis Web site at
http://www.taylorandfrancis.com

and the CRC Press Web site at
http://www.crcpress.com

Contents

Preface

What does it mean, "to do mathematics"?

"Let me put up a few numbers to make the discussion more concrete" is a wonderful sentence that all regular talk-show economists should have in their arsenal. I even wanted to use it as the title of this book, but my publisher countered with the argument (though it seemed specious to me) that it was too long for a book title. We compromised on the shorter version you see on this book.

Put up some numbers—and then what? Then they are up. Just so.

I could also have called the book *What Is Mathematics?* but that is already the title of a book by Richard Courant and Herbert Robbins, a book about which one could say a few things. For example, that Courant selected the rather catchy title on the recommendation of Thomas Mann. Aside from the fact that Mann is no longer available to help with my title selection, I also did not want to write a classic mathematics book like the one by Courant and Robbins, in which mathematics is displayed along with mathematical concepts, ideas, considerations, research, and results.

My book is to be about the *doing* of mathematics, the *making* of mathematics. That is something entirely different. Try saying "love" instead of "mathematics." So what is love? What is making love? Of the latter we have some pretty concrete

ideas, even if, in German, is comes across as a bad translation.
My book is to be about the people behind the numbers and
the places where mathematics is made. It should be about the
battles for precision; about perseverance, errors, and the love
of detail; about great emotions—and also about the problems
that make the fight worthwhile, about recognition and prizes.
This book is a trip into the world of mathematicians. This is
no separate secret world. The world of mathematics is *our*
world. Mathematics is not at all something distant, strange,
and abstract that one can only learn about, and learn to hate,
in school. Instead, it designates the corner pennants and goal
lines of a playing field across which we move quite unen-
cumbered, a field that we don't first encounter when we learn
how to count, one, two, three. Are you aware that mathematics accompanies us from the very first knot in a shoelace until
the artificial knee? There is mathematics in housekeeping, in
communications, in traffic, and in weather reports (especially
when they are accurate).

The world of mathematicians is also nothing foreign. You
will find in this book all the rubrics familiar from magazines
and daily papers, because they are interesting: celebrities,
history, travels, politics, science and technology, weather,
clever puzzles, a look into the future, and not a lot of eso-
terica. For which points to emphasize, I have of course
oriented myself toward my own interests, preferences, and
aversions by painting, in a manner of speaking, a picture
of this fascinating and far-reaching world of mathematics
and the making of mathematics *as I see it*. Think of it as an
adventure trip with a personal tour guide. In that sense: welcome aboard!

Chapter 1

On the Number Line

"All is number," the motto of the Pythagoreans, designates the belief that the regular behavior of the world can be grasped and expressed with numbers. We still believe that today, and not without reason.

But if numbers are so fundamental, then we have to allow ourselves to ask, "What are numbers, and what are they good for?" The question may seem stupid or naive, but it is neither. The number theorist Richard Dedekind formulated the question in this way as the title of his famous book *Was sind und was sollen die Zahlen?* And the question has no simple answer. As indeed it cannot; it may well be that the regularities in the world can be phrased and understood in terms of numbers, but if so, then "one, two, three" will certainly not suffice. In fact, the so-called *natural* numbers (one, two, three, and so forth), which seem to us so concrete and obvious, already present problems—philosophical problems, of course, but also very concrete problems.

Therefore, once again, what are numbers? Are they something that designates a quantity? In that case, is $\frac{1}{2}$ a number? Or –1? Or $\sqrt{2}$? Something with which one can count? In that case, is "infinity" a number? Something with which one can compute? A domain in which one can solve equations? In that case, the "imaginary unit" $i = \sqrt{-1}$ is a number. And, as if the question were not already unclear enough, it seems as if mathematicians can't get enough, they are always creating new numbers—or will they at some point finally be satisfied?

3—Can Bees Count?

In January 2009, many newspapers around the world reported the news that "Bees can count to three." Researchers at the Bee Group of the University of Würzburg were said to have determined that bees could indeed count to three.

However, I also remembered seeing a headline not too long before that bees can count to four. What was going on? In this case it was not my bad memory, since the much more complete and thorough memory of Google confirmed that "Bees can count to four" was reported in October 2008 by *Netzeitung* (www.netzeitung.de), for example.

Let us leave aside for the moment that we find it extremely surprising that bees can count at all, never mind whether it is up to three or up to four. And let us leave aside any mysterious auras that surround these numbers. Three, like each of the smaller integers, is freighted with all sorts of symbolism, including, for example, the Christian Trinity, which involves the belief that the Father, Son, and Holy Ghost are together *one* God, or in simple terms, "one equals three." This opens up all sorts of avenues for further discussions, but that will not be our theme. Three is for us nothing but a number—but what does that mean?

The two news reports were referring to different experiments by different teams of bee researchers. In the first experiment, bees were trained to fly toward panels that showed objects, where they were rewarded with sugar water. The bees learned that it was only panels with three objects that led to food, not panels with four or six objects. They were able to make this distinction whether the panels had apples, flowers, or red or black spots. The researchers concluded that the bees had been able to form an "abstract" notion of the number 3 and could differentiate it from 4. They could, after training, fly toward panels with three objects instead of panels with four, five, or six objects. The bees could, however, not be trained to prefer panels with four objects to those with five objects, which led the researchers to conclude that bees cannot distinguish 4 from 5 nor 5 from 6. Thus the report that bees can count to three. Quite an achievement for a little animal with a brain the size of a sesame seed. Chimpanzees and humans can recognize four objects at a glance, but no more. With five or six objects it has been shown not to be possible to see "at a glance" how many there are; for that many objects, one has to start counting.

Do we now have to imagine the clever bee Maya and her somewhat duller friend Willy who mutter "one, two, three…" to themselves and point their fingers in the air? But we know bees don't have fingers. And they probably mumble only in animated television series.

As for the second report, to make bees count, one could try the following experiment: let bees fly through a tube with marks on the side and try to train them to look for food after the third mark. The marks could be put at different distances each time, so that the bees would be prevented from looking for the food at some particular distance. And indeed, the bees can learn to count to three, that is, to fly to the third marker. They can also be trained to count to four, that is, to fly to the fourth marker. But farther than that they cannot count, not even after very patient training.

Thus it came to be that in German newspapers it was reported that "bees can count to four"—and perhaps raising the question of whether journalists can count at all in a reader's mind a few weeks later when one saw the headline "Bees Can Count to Three." In any case, Professor Srinivasan, one of the investigators, reports that there are slower learners and faster learners among the bees, that is, cleverer and duller bees. But we knew that already, and we like Willy anyway.

So, do the bees then know what the number 3 *really* is? That is a fruitful philosophical question that we cannot leave entirely to the bees. However, among the reliable and firm foundations of mathematics is, naturally, an expectation that mathematicians will have neat and clear answers and concepts for such questions. They have them, too, but not for as long as one might think: this is not one of the questions answered in the depths of time or even in some Greek dialog from classic times. It was only toward the end of the nineteenth century that Georg Cantor clarified the difference between cardinal and ordinal numbers in the course of his investigations of set theory. The former describes the size of a set (a set can have one, two, three, or more elements; one is concerned with the quantity). The latter arises in the sorting of elements and then in counting them off (in which position is an element in a sequence?). There is an enormous difference, even when bees are counting. Only the journalists (and the headline writers) missed the distinction. It is, after all, not simple.

Even so, should we be impressed that bees can count to four? Actually, no. Much more impressive is the "waggle dance" in which bees dance in a geometric pattern to tell their hive mates the location of food sources. It's a powerful dance: the angle from the vertical of the line along which the bee waggles indicates the angle from the direction to the sun in which the bees must fly to find the buffet. Clearly, bees are inclined more to geometry than to arithmetic. As you can see, mathematics is rich and varied enough that all can use their talents.

5—Can Chickens Compute?

Another one of these headlines that seems to undermine our presumed supremacy in the realm of mathematics: "Chicks Can Compute—At Least Up to Five." An Italian researcher named

Rosa Rugani and her colleagues found this out. The news was circulated around the world on April 1, 2009, occasionally with an explicit comment that this was *not* an April Fool's joke. The BBC website published the result with the provocative headline "Baby Chicks Do Basic Arithmetic" and added appropriately adorable pictures of baby chicks. The scientific publication supporting the claim, according to the report, appeared in the renowned *Proceedings of the Royal Society B: Biological Sciences.*

In fact, I was able to find, with help of Google, the appropriate page, also published online on April 1. There it is claimed that freshly hatched chicks can compute, for example, "2 + 3 = 5" in their heads (where else?), which Rosa Rugani was able to confirm through a tricky set of experiments.

I was skeptical—and disturbed. The little chicks can do math? Is this an April Fool's joke after all? Suppose we take it seriously, which would mean that at least primitive arithmetic operations such as addition are not only child's play but even something freshly hatched chicks can do. But what would the little fuzzy beasts gain thereby? Some evolutionary advantage? I am definitely of the belief that intelligence has brought us (!) some (?) evolutionary advantages (although this is far from proven)—but chicks?

So I wrote on my blog "Mathematik im Alltag" (www.wissenlogs.de) with a headline "Chicks Can Compute? Help!" and asked readers for clarification and explanation. The first response was from a reader who was quite certain that Ms. Rugani conducted only serious research that was contributing further proof that chickens are smarter than we think. The reader's name was Martin Huhn (but if my blog were in English, he probably would have signed himself Martin Chicken).

And then I remembered that already in the 1980s a certain Luigi Malerba from Italy had reported:

A learned hen wanted to teach her colleagues to count and to add. So she wrote the numbers 1 through 9 onto one of the walls of the chicken coop and explained that one can get even larger numbers by combining these. To teach the others addition, she wrote on the next wall: $1 + 1 = 11$, $2 + 2 = 22$, $3 + 3 = 33$, and so forth, until $9 + 9 = 99$. The hens learned to add and found it quite useful.

This clarifies everything.

10—And the Name of the Rose

A friend, and proud father, recently told me that his two-year-old son could already count to five. Only, he does not like two, so he counts, "one, (short pause), three, four, five." Now we have to clarify just what he means by "three," and we better do it soon, before he *really* learns to count.

The same problem occurs elsewhere: most airplanes have no row to which the label "13" is attached; of course, the passengers in the row labeled "14" are nonetheless sitting in the thirteenth row (and we hope they feel safer because the row is mislabeled). In many theaters and opera houses, the pleasure of having cadged a seat in the first row evaporates when one finds that one is sitting seven rows from the stage, with heads from rows A, B, C, etc., or even AA, BB, ... blocking the view. (Sometimes this is a good thing—for example, if the stage is so high that the people sitting in what is really the first row need chiropractic assistance after the performance.) Or think of Douglas Adams's science fiction trilogy *Hitchhiker's Guide to the Galaxy*. The cover of the fifth (!) volume included a note that this is a book that "gives an entirely new meaning to the concept of trilogy." That it does.

The redefinition of numbers is an everyday phenomenon that we trip over everywhere. Nonetheless, the father's concern about his son's "wrong counting" is noteworthy. Of course, one could count differently, that is, use different words for the numbers. After all, we have an unmovable notion of the number 7 and of "counting to 7" that is completely independent of what we call the number, say, ☺ ☺ ☺ ☺ ☺ ☺ ☺. But why do we name the numbers as we do? And are the names we give the numbers at all important? "What's in a name? A rose by any other name would smell as sweet," says Juliet about her Romeo. Is that true for numbers, too? A brief look at history at least shows us that our numbering scheme, the Indo-Arabic positional notation with base 10 (and an additional complication arising from an occasional switching of numbers when speaking) is not at all self-evident and indisputable and not without alternatives.

The Story of the Zero

The "discovery of zero" seems like a little trifle, but it is in fact an important achievement of civilization with implications perhaps as dramatic as the discovery of the Americas (by Columbus in 1492) or of penicillin (by Alexander Fleming in 1928), even if we don't know who discovered zero or when. In fact, zero was apparently discovered at least three times in the course of history, and each time with a different meaning: the Babylonians around 700 BCE used a symbol of three ticks as a place holder; the Olmecs and, later, the Maya in Central America used (long before Columbus) a zero mark in their calendar and later in their base-20 number system; and, lastly, in India in the fifth century. There the zero was used not only as a numeral and as a placeholder but also as a number with which one could *compute*. Accordingly we must thank some anonymous Indian scholar for our positional number system, in which "2001" designates a number that is the sum of two thousands, zero hundreds, zero tens, and one unit—a very different number from 201 and 21.

This Indian positional number system subsequently came (with modifications of the symbols used for the numerals, but without modification of the principle) to Europe via the Arab–Islamic trading routes. According to legend, the mathematician Gerbert de Aurillac (c. 945–1003) played an important role in the transfer after he became Pope Sylvester II in 999, but that is unlikely, because the pope, like everyone else in Europe at the time, used an abacus to crunch his accounts. In the twelfth century, the Indo-Arabic number system first entered Western Europe through the translation of an Arabic book on computation. Widespread adoption, however, came only after the publication in 1202 of the influential *Liber Abaci* by Leonardo of Pisa, also known as Fibonacci (c. 1170–1240). Among the "common folk" in Germany, the base-10 system, including zero, became widely accepted only at the beginning of the sixteenth century. Adam Ries (c. 1492–1559) ran a school for computing in Saxony, which was subsequently continued by his sons. His second textbook, *Rechnung auff linihen und federn* (*Computation with Lines and Feathers*), taught computation not only with an abacus ("lines") but also with Indo-Arabic numerals (written with "feathers," that is, pens). The book was a remarkable bestseller, with at least 120 editions, and was used as a textbook well into the seventeenth century. Ries's name has become a byword for Germans who "know their math," for whom 120 + 69 makes, "from Adam Ries," 189.

The positional system allows us to write large numbers without much effort—and certainly far less effort than the Roman numerals it replaced. Although using MMI instead of 2001 seems like a minimal effort, the federal budget, with its millions and billions of dollars, would require a lot of ink: a billion is a million thousands, which means the Romans would have had to write it as:

MMM MMM MMM MMM MMM MMM MMM MMM MMM MMM MMM MMM

with a million Ms written out. Not very practical.

Apropos of large numbers: when it was discovered that a minor functionary in a French bank had gambled €50 billion and pilfered a few for himself, my daily newspaper, the Berlin *Tagesspiegel*, reported that "if one wanted to write this unbelievably large number here, the last zero would presumably be somewhere in the classifieds." By which was meant, presumably, that the number for 50 billion would require some fifteen pages to write out. Of course that's not the case, not even if one uses Roman numerals. Apparently the *Tagesspiegel* has difficulties with large numbers. But this is a tradition. There have been finance ministers in Germany who were hard-pressed to say how many zeroes a billion actually has. Even in the course of an election campaign, the finance minister then in office managed to confuse billions with trillions while the TV cameras were on.

Base 10

Of course, ten fingers. Never mind that I know mathematicians with nine fingers. One could just as easily ignore the thumbs and count with a basis of 8 or of 4. And in fact, there are Indians in South America, for example, who do just that. Or one could include fingers and toes, for a basis of 20. Even base 12 was once widespread; its vestiges can still be found in our counts of hours and months. The Babylonians computed with a base of 60. (This is impractical for computations: the "small" multiplication table is impossibly large!) Electronic computers use a binary notation, that is, a basis of 2, which is practical for them. The nicest numerals I know were designed for a base-8 system, called "octomatics." The numerals are not 0, 1, 2, etc., but much more systematic: _____ for zero, _____⌐ for one, __⌐__ for two, and then ⌐⌐, ⌐___, ___⌐, ⌐⌐_, _⌐⌐, for three through seven. This is easy to understand and easy to remember (a line on the right is one, one in the middle is two, and one on the left is four, and then one adds); the numerals are easy to write and easy to compute with, but I doubt that they will be widely adopted.

Turned Around

It would be logical to pronounce the Indo-Arabic numerals in the same way one writes them, that is, from left to right. Thus 3213 would then be three thousand two hundred ten and three, just as 3123 is three thousand one hundred twenty-three. That would be logical, but it is not done. In some languages—German, for example—this reversal of ones and tens is carried throughout: 51 is one and fifty, 98 is eight and ninety. This reversal of the digits is quite old, dating to Indo-European linguistic roots, and has persisted quite stubbornly. In the renowned textbooks of Mr. Ries, however, one was instructed to pronounce the numbers from left to right; no mention was made about correcting the reversal of the last two digits. Another German author, Jakob Köbel, town clerk of Oppenheim, included a pronunciation table of the numbers from 21 to 99 in the left-to-right order, but without commenting that this is unusual for German; he is, however, inconsistent, just in the way English usage is inconsistent: the numbers 13 through 19 are *not* ten-three (*zehn-drei*), ten-four (*zehn-vier*), and so on, but thirteen (*dreizehn*), fourteen (*vierzehn*), and so forth.

Even if we admit that it is more logical, more consistent, simpler, and less likely to lead to mistakes, to say all numbers—including one through twenty, or, in German, one through ninety-nine—in the same order as we write the digits, we can also see that this will probably not happen in either English or German. Other peoples are more willing to reform—for example, the Norwegians, whose pronunciation of numbers was reformed by law in 1950. In Germany there is an enterprising society named "Zwanzig-eins e.V." ("Twenty-One,

Inc."), led by Lothar Gerritzen, professor emeritus of mathematics at the University of Bochum, that has been trying to persuade schools to admit the unreversed pronunciation (*zwanzig-eins, zwanzig-zwei*) at least as an alternative. I think this is correct and desirable. It would encourage a bit of fantasy in dealing with numbers, as well as a vitally necessary bit of anarchy. It might even be comparable to the introduction of the revised German orthography, which simplified the rules for spelling German words in the 1990s and which mostly had the effect of replacing the old standard with whatever the writer thinks the new correct spelling might be. This kind of variety and anarchy is surely to be welcomed.

On the other hand, the reversed pronunciation of numbers between eleven and ninety-nine is a part of German culture, something that no law can change. In English it also remains a vestigial and poetical part of the language (and not just vestigial for the numbers less than twenty): we will always have "four and twenty blackbirds baked into a pie," even as the Beatles sing "When I'm sixty-four."

13—Bad Luck?

The number 13 is probably the number that is most freighted with emotional baggage. It portends bad luck. Friday the 13th is particularly dangerous. We all know that. It is nonetheless noteworthy. Apropos of emotions: do you like 28? How do you feel about 9973?

Daniel Tammet is the young, autistic British author of *Born on a Blue Day: A Memoir of Asperger's and an Extraordinary Mind* (the title of the German edition is *Elf ist Freundlich und Fünf ist Laut,* or *Eleven Is Friendly and Five Is Loud*). For Tammet, 13 is a small number. It is also a prime number, and for Tammet, primes have "smooth and round shapes, similar to pebbles on a beach." Otherwise, according to Tammet,

13 is nothing extraordinary, and certainly nothing alarming. Numbers, Tammet says, are his friends, and one does not consider friends en masse; one celebrates their individuality. So just what makes 13 so special? Mathematics does not help. Yes, 13 is a prime, but one of infinitely many primes. It is also a Fibonacci number, encountered in the sequence 1, 1, 2, 3, 5, 8, 13, 21, 34, …, in which each number is the sum of the two preceding numbers. Whether there are infinitely many primes that are also Fibonacci numbers is not known—an unsolved problem, but not one that makes 13 special in any mathematical way.

In Italy, where it's actually 17 that is the unlucky number and 13 is considered a lucky number, it is nonetheless considered a fatal error to have thirteen people sitting at a meal—perhaps as reverence for the Last Supper. That belief runs sufficiently deep that Prime Minister Silvio Berlusconi is said to have once banished Sandro Bondi, a member of Parliament (and later Berlusconi's minister of culture), to the kitchen when he realized that there would be twelve guests at his dinner table.

That Friday the 13th is a big deal is a modern invention that first arose in the late nineteenth century. We can even surmise the identity of the originator: an American named Thomas William Lawson, who made a fortune speculating on the stock market. In 1907 Lawson published a novel about the market called *Friday the Thirteenth* in which a broker picks that day on which to bring down Wall Street; it was published in a German translation that same year, and somewhat later it was made into a film. The only seven-masted schooner ever built was named after Lawson; it was wrecked on Friday the 13th, 1907. Of course, once a juicy theme such as the connection between Friday the 13th and bad luck was brought up, it easily reinforces itself. So it hardly matters that the stock market crash on "Black Friday" in October 1929, which led to the Great Depression, actually happened on Thursday and continued on the following Monday, and that it was the 24th of October and not the 13th. But maybe that's a hallmark of the unlucky 13, that it hides behind other numbers.

Popular wisdom also knows that there are more traffic accidents on Friday the 13th than on other days. The explanation is easy: people drive less well on a Friday the 13th because they are more nervous, and, bang, an accident. However, whether this is in fact correct and whether it can be statistically corroborated is not clear. According to an analysis by Edgar Wunder published in 2003 in the *Zeitschrift für Anomalistik* (or, *Journal for Anomalous Phenomena*, a journal I had not known about and which apparently deals skeptically with skeptics), there are statistically no more and no fewer accidents on a Friday the 13th than on the preceding Friday the 6th or the subsequent Friday the 20th.

Conspiracies? Self-fulfilling prophecies? In any case, the morbid fear of the number (or, rather, numeral) 13 is so widespread and so serious, that psychiatry has a term for it: triskaidekaphobia. The word is easier to pronounce if you know a little Greek or if you have no fear of foreign languages (for which there is surely also a pseudo-Greek technical term). Even harder to pronounce is paraskavedekatriaphobia, a fear of Friday the 13th.

Is the number in any way notable—other than the psychological effects, which, after all, say little about the number itself? As a lottery number, 13 appears relatively seldom, at least thus far in the German state Lotto "6 out of 49," and then only when we restrict it to Saturday drawings (on Wednesdays, 32 is particularly rare, and recently also 21). However, the differences are not statistically significant, and you can be sure that this has been investigated carefully. There is thus no basis for conspiracy theories ("the ball with '13' is a little smaller than the rest").

On the other hand, in our calendars, the 13th of the month is especially often a Friday. This is a peculiar result of the Gregorian calendar reform, which was introduced to Roman

Catholic countries in 1582 by a papal bull and has since then spread throughout the world. The Gregorian calendar has 365 days per year, except for years divisible by 4, which are leap years with 366 days; years divisible by 100 are usually not leap years, but years divisible by 400 are. As a result, the Gregorian calendar repeats itself every 400 years. This cycle has 400 × 365 days plus 100 leap days minus the three for the exceptions in centuries. That makes in all 146,097 days, which is divisible by seven, making exactly 20,871 weeks. Thus, for example, May 19, 1963, and May 19, 2363 (my 0th and 400th birthdays) fall on the same day of the week.

The 400 years are also 4800 months, and so there are precisely 4800 dates that are the 13th of the month. But 4800 is not divisible by seven, and therefore it is not possible for each day of the month, such as the 13th, to fall equally on each day of the week. A direct count shows that the 13th falls most often on a Friday (688 times) and least often on a Thursday or a Saturday (684 times).

Is this an accident? Of course not. It depends on which day of the week the calendar was begun, in other words, on the precise details of how the Gregorian calendar was introduced—all, of course, blessed by the pope. To keep the date of the spring equinox around the 21st of March, ten days were dropped from the year 1582, so that Thursday the 4th of October 1582 was immediately followed by Friday the 15th of October. This was an arbitrary decision by the pope. Had he instead decided to follow Thursday the 4th with, say, Monday the 15th, then the 13th days of the month would most often be a Monday. Because of the papal reckoning, the 13th is not only most often a Friday, but the months when the 13th is a Friday appear in clusters. For example, if there is a Friday the 13th in February (not in a leap year), then there is another again immediately in March, and another in the following November. We had this in 2009, and it will happen again in 2015. But we know that misfortunes rarely come alone.

(I wrote this chapter on the 8th of April 2009. I had thus survived two Fridays the 13th that year already. The third did not find me unprepared.)

42—The Answer to Everything?

The number 42 is the answer to the Ultimate Question of Life, the Universe, and Everything. This is one of the first things one learns in the first volume of the trilogy *The Hitchhiker's Guide to the Galaxy* by Douglas Adams. The answer 42 was the result of seven and a half million years of computational effort by a computer named Deep Thought, built by a race of hyperintelligent mice; unfortunately, however, it was not clear what the *question* was to which 42 is the answer. As a result, a second, even larger, and more complicated computer had to be built to determine what the question was. This computer was called Earth; sadly it was destroyed before it could conclude its task.

Could a number, a totally banal number like 42, be the answer to the Ultimate Question of Life, the Universe, and Everything? That's certainly food for thought. But it gets worse. At the end of the second volume, we learn that the question was actually, "What is six times nine?"

Before you object that "Hey, that's not 42!" please recall that, while we all believe that mathematics is an internally consistent system, and that, in particular, operations with the natural numbers 1, 2, 3, …, lead to no contradictions (as long as we make no computational mistakes), this has not been proven. Indeed, even though all mathematicians rely on it, it is not proven—and not even provable.

Not that mathematicians haven't tried to prove the internal consistency.

At the end of the nineteenth century, it was discovered that one could, even in mathematics, produce self-referential sentences that cannot be either true or false, much like the statement of a barber that he shaves all the men in his village who do not shave themselves. (What of the barber? If he shaves himself, his statement is incorrect, for then, as he said, he would not shave himself; but if not, then he would....) Mathematics, of course, does not have to deal with barbers, no matter whom they may or may not shave. However, mathematicians do deal with sets and must therefore be concerned with such constructions as "the set of all sets that do not contain themselves as an element"; whether that set contains itself can be neither true nor false as either leads to a contradiction. This conundrum led to a deep crisis in the foundations of mathematics. To overcome it, David Hilbert, a mathematician at the University of Göttingen, proposed an ambitious program that was to set a firm foundation for all of mathematics. The program foundered unexpectedly on the shoals of a theorem formulated and proved (!) by Kurt Gödel that states: it is impossible to prove that elementary arithmetic—that is, the computation with natural numbers—is free of contradictions. Any finite set of axioms one tries to use to establish an internal consistency for arithmetic is incomplete.

Thus mathematicians are somewhat helpless when Adams's computer asserts that $6 \times 9 = 42$, a statement that serves as evidence that "something is fishy with the world." Perhaps we all suspect, deep down, that there is indeed a lot fishy with the world, but we hope that at least the basics are nonetheless firm. Among those basics, we should include (along with free will and the rules of the road) the computation with natural numbers.

Gödel's Incompleteness Theorem opened a door not only for the British author Douglas Adams but also for the Swedish philosopher Astrid Lindgren, whom we have to thank for the following lines

> Two times three is four
> Tjolahey, tjolahopp, and three makes nine
> I make my world
> Tjolahey, the way I like

(loosely translated from the German version of the title song for the 1968 movie *Pippi Longstocking*), immediately followed by the offer:

> Three times three is six
> Tjolahopp, who wants to learn from me,

as well as:

> And anyone who likes us
> Will have to learn our way to count.

Let's just assume that the song does not really point to problems in the foundations of arithmetic but rather is an expression of considerable imagination and occasional computational errors. Could we, then, propose Saint Pippi, whose full name is Pippilotta Delicatessa Windowshade Mackrelmint Ephraim's Daughter Longstocking, and who is the inventor of "plutimication," as the patron saint of arithmetic errors? Perhaps we could obtain testimonials for her cause from powerful advocates, including one or another finance minister, whose deficit computations look more and more like the result of a division by zero.

On the other hand, perhaps an even higher authority on errors in arithmetic is Microsoft, whose Excel 2007 program cheerfully computes:

$$850 \times 77.1 = 100,000$$

(the conventionally correct answer is 65,535). Is this better or worse than $6 \times 9 = 42$ or $2 \times 3 = 4$? Once again, one loses a little faith in (choose one) the meaning of life or programs for computers.

91—The Numbers on the Bone

The prime numbers are the atoms of arithmetic: indivisible numbers. More precisely, they are natural numbers that one cannot write as products of smaller natural numbers: 2, 3, 5, 7, and so forth. (By convention, 1 is not a prime number.) Prime numbers are also at the start of mathematics because very quickly after the operation of counting, multiplication and, of course, division come next.

How quickly this happens is shown by a piece of pre-historic evidence. The Natural History Museum in Brussels preserves a small, fossilized animal bone, only ten centimeters long, that was found in 1950 in Central Africa, at Lake Ishango in the Congo, near the border with Uganda. The bone was found in the remains of a lakeside settlement that was buried by a volcanic eruption—a sort of prehistoric Pompeii. Modern radiocarbon dating indicates the bone is at least 20,000 years old. As such, it is nothing special. But it is, in fact, unique and quite noteworthy. At one end it is decorated with a crystal, and thus may well have had ritualistic significance. But this would not make it unique. What is significant is that the bone has little grooves arranged in groups and lined up in three rows. In one of these rows the groups have 9, 11, 19, and 21 grooves (altogether 60 grooves); in one of the other rows the groups have 11, 13, 17, and 19 grooves—these are the primes between 10 and 20, and their sum is also 60.

Is this just an accident, that this small bone shows the prime numbers between 10 and 20? I don't know. Nobody knows. Perhaps 11, 13, 17, and 19 are significant if one uses a base-6 numbering system, because these are multiples of 6, plus or minus one. In any case, 20,000 years ago, when the inhabitants of the shores of Lake Ishango had no writing—indeed, long before the invention of writing—someone played around with numbers and came across prime numbers, the atoms of arithmetic.

The groups of grooves on the bone from Lake Ishango are the oldest known record of a culture with a mathematical awareness. Amazingly, it shows numbers that could not have been fully understood by the people who made the marks—since it is extremely unlikely that the people at Lake Ishango could multiply and divide (or is that a lack of imagination on our part?). If one cannot multiply, one cannot determine whether the numbers are primes. A number that is not a prime is composite, that is, it has factors, which themselves must be either primes or products of other factors. For example, consider the number 91. It is smaller than $100 = 10 \times 10$; if it has factors, at least one must be smaller than 10—if both factors were larger than 10, the product would have at least three digits. A two-digit number such as 91 is thus a prime number if it is not divisible by 2, 3, 5, or 7. Numbers divisible by 2—that is, even numbers—are easily recognized, since their final digits must be even (2, 4, 6, or 8); multiples of 5 are also easily recognized by their final digits (0 or 5); numbers divisible by 3 can be recognized by the fact that the sums of their digits is also divisible by 3. Since 91 has none of these properties, the remaining possible factor is 7, where there is no simple way to see if it is a factor; 91 could, of course, also be a prime.

That's why I like the number 91, my "magical number": it looks very much like a prime number, but it isn't one: $91 = 7 \times 13$.

Today we know a lot about prime numbers, but we still do not understand them completely. We can determine relatively easily whether a number is prime—both theoretically and practically. The theoretical problem is one of the important mathematical problems that have recently been solved. Carl Friedrich Gauss formulated it as follows (in his major work, the *Disquisitiones Arithmeticae*, of 1801):

> That the task of distinguishing the prime numbers
> from composite numbers and of decomposing the
> latter into their prime factors is one of the most
> important and useful tasks of all of arithmetic, and
> that this task has occupied the minds and efforts of

the older and younger geometers is so well known
that it is superfluous to lose any words about this
topic here.

In fact, Gauss is here conflating two separate tasks: first, to
"distinguish" the prime numbers from composite numbers, and
second, to "decompose" the composites, that is, to factorize
them. The second task is still considered difficult and is probably not possible in practice. The first, however, is solved. On
August 4, 2002, Manindra Agrawal of the Indian Institute of
Technology in Kanpur and his undergraduate students Neeraj
Kayal and Nitin Saxena sent their solution to fifteen experts.
Four days later, the *New York Times* ran an article with the
headline: "New Method Said to Solve Key Problem in Math." In
2004, the solution appeared as a paper in the journal *Annals
of Mathematics*. With the AKS algorithm (named after Agrawal,
Kaya, and Saxena), one can efficiently and with 100% certainty
determine whether a number is prime. Of course, this is effi-
ciency in a theoretical sense. In practice, if one needs to determine rapidly if a particular number is prime, one will use not
the AKS algorithm but rather a set of well-established tricks
and shortcuts, which may in theory not work on occasion, but
which in fact do impressively well: a typical laptop computer
can check the primality of a number with several hundred digits without too much effort.

Gauss's second task, the actual decomposition of composite
numbers, is considered difficult. The British economist and
logician William S. Jevons observed already in 1874 that while
multiplication is easy, factorization is not:

Can the reader say which numbers multiplied
together give the number 8,626,460,799? I hold it as
highly unlikely that anyone other than me will ever
know the answer.

Actually, we now know the answer. The number theorist Derrick Norman Lehmer solved the problem in 1903. Today, the computer algebra program Maple readily solves the problem on a PC with a few keystrokes: 96,079 × 89,681. Nonetheless, Jevons is right. It is easy to multiply two numbers, no matter how large; it is easy to find large primes and to certify that they are prime, but factorizing a 200-digit composite number is still difficult, even using a supercomputer.

Jevons's considerations were not just an academic game. He wanted to use them for constructing codes, that is, to go from "multiplication is easy, factorization is hard" to "encoding is easy, decoding is hard." As it turns out, Jevons's thoughts were an early precursor to the so-called RSA encoding scheme, which was proposed in 1977 by Ron Rivest, Adi Shamir, and Leonard Adelman (all at the Massachusetts Institute of Technology). In 1982, the three MIT mathematicians founded a company, RSA Data Security, and obtained a U.S. patent for their procedure a year later, and we can hope they made a lot of money with it. Variations on the RSA algorithm are still in use today for encoding transmissions on the Internet, such as banking transactions.

A great deal of the security on the Internet thus depends on the difficulty of factorizing composites—even if we already know that they are composites. If someone could devise a method for factorizing large composite numbers (e.g., with 400 digits), they could read all sorts of messages on the Internet that had been thought to be private as well as manipulate banking transactions and other businesses. Clearly, many organizations, including intelligence services, are very interested in the problem. In the United States, the National Security Agency employs several hundred mathematicians, including number theorists, to study encoding and decoding schemes. Even the RSA algorithm was known to some intelligence services, including the British Government Communications Headquarters, GCHQ, long before 1977.

What they really knew, and what they could do, I don't know, and no one knows exactly—and that's true for the mathematicians employed by the intelligence services just as much as the prehistoric people on Lake Ishango some 20,000 years ago.

1729—Hardy's Taxi

The smallest prime number is 2. The smallest odd prime number is 3 (and bees can count to 3—or is it 4?). The smallest composite number is 4; it is also the smallest number that is the sum of two primes. Next is 5, which is a prime, but also the sum of two primes. (An unsolved problem: are there infinitely many primes that are also the sums of primes? One suspects there are.) Since $5 = 2^2 + 1$, 5 is also a Fermat prime, that is, a prime number that is two raised to some power plus one. (Another unsolved problem: are there infinitely many Fermat primes? Presumably not.) Next, 6 is a "perfect" number, that is, it is equal to the sum of its factors: $6 = 1 + 2 + 3$. (Unsolved problem: are there any odd perfect numbers?) The number 7 is a Mersenne prime, that is, two raised to power of a prime number minus 1. (Unsolved problem: are there infinitely many Mersenne primes? Presumably yes.) It is also the smallest number n for which an n-sided polygon cannot be constructed with just ruler and compass. (This has been proven.) The numbers 8 and 9 are interesting because they are the only two neighboring numbers both of which are powers of other numbers: $8 = 2^3$ and $9 = 3^2$. (This was conjectured by Eugène Catalan in 1844 and thus called Catalan's Conjecture; it was proven by Preda Mihăilescu in 2002.) Of course, 10 is interesting because it is the base of our number system; 11 is one of a pair of prime number twins (11 and 13); 12 has especially many divisors; 13 is unlucky or, perhaps, not; and so forth. And so forth? Does every number have a story, a peculiar property, an interesting construction, or some con-jecture associated with it? Are all numbers interesting? Well, if not all numbers are interesting, then there must be a smallest uninteresting natural number. However, that leads to a logical contradiction, because the smallest uninteresting number will necessarily be interesting because of that fact. Logically, there-fore, there can be no smallest uninteresting number, which is thus a proof that all numbers are interesting.

Where is the error in this argument? It probably lies in the fact that "interesting" is not a well-defined mathematical prop-erty but rather a matter of taste and preference. And about preferences one should not argue, and interests are not easily nailed down. After all, numbers could be not only interesting or uninteresting, but also very interesting or less interesting, mildly uninteresting or quite uninteresting.

Aside from that, it is also a question of mathematical insight to recognize interesting details about a number. The British number theoretician G. H. Hardy is said to have visited his colleague and protégé Srinivasa Ramanujan in the hospital. In the course of the conversation, Hardy mentioned that he had come to the hospital in a taxi numbered 1729, a rather unin-teresting number. "No!" answered Ramanujan, the number 1729 is quite special since it is the smallest number that can be written in two different ways as the sum of two cubes: $1729 = 9^3 + 10^3 = 1^3 + 12^3$.

People interested in history can also find 1729 notewor-thy because it is the year in which J. S. Bach's *St. Matthew Passion* was first performed and in which Catherine II, tsarina of Russia, was born. But do these accidents of our calendar make the number 1729 interesting? Let me add something from mathematics: 1729 is a number that looks like a prime, but is not, for it is divisible by $13 = 1 + 12$ and by $19 = 10 + 9$; and this is no accident but is connected with Ramanujan's insight.

119/100—High Percentages

Do you have a favorite fraction? It is likely that no one has ever asked you that. A favorite number, perhaps, but a favorite fraction?

I have asked a few colleagues that question and was mostly rewarded with strange looks (perhaps not wanting to say out loud, "a little odd, sometimes, that Ziegler"). Most mathematicians when asked for a favorite number will say, 7, or sometimes 42, or perhaps π, the ratio of a circle's circumference to its diameter.

Only my colleague Christof Schütte of the Freie Universität Berlin, whom I pestered with this question at a garden party celebrating his forty-third birthday, answered without hesitation that it was "$\frac{3}{7}$, and always has been." Why was that? He couldn't say exactly, but he always liked the decimal expansion of $\frac{1}{7}$, which repeats itself after every sixth digit:

$$0.142\ 857\ 142\ 857\ 142\ 857\ 142\ 857\ 142\ 857\ 142\ 857\ 142\ 857\ 142\ 857\ 142\ 857$$

and the first two places after the decimal point are the same as those of π. In addition, the multiples of $\frac{1}{7}$ all have the same repeating sequence, just starting at a different position, so,

$$\frac{3}{7} = 0.428\ 571\ 428\ 571\ 428\ 571\ 428\ 571\ 428\ 571\ 428\ 571\ 428\ 571\ 428\ 571\ 42$$

and it even starts with 42!

Nonetheless, I would argue that the most important fraction in my life—and of everyone who lives in Germany—is not 3/7 but 119/100 = 1.19. It is a number that enters nearly all of our commercial transactions, because the multiplication by this fraction is what is commonly referred to as "including the value-added tax." (This is not universal, however. In New York City, which I visit occasionally, the fraction

100,875/100,000 serves a similar function by adding a sales tax to the transaction.) Of course, 119/100 is not for this reason anyone's "favorite" fraction; only in the Finance Ministry are there people who might get anything like enthusiastic about it.

π—As Beautiful as the Mona Lisa?

Both religion and mathematics claim to provide eternal truths—however, to the right of the decimal point, mathematics is considerably more precise. As evidence one can turn to the Old Testament, first book of Kings, Chapter 7. There one finds a report of the construction of the temple by Solomon. To furnish the temple, Solomon called upon Hiram of Tyre, who was "a worker in brass; and he was filled with wisdom, and understanding, and cunning to work in all works in brass." Hiram made two pillars of brass for the porch of the temple,

> And he made a molten sea, ten cubits from the one brim to the other: it was round all about and his height was five cubits; and a line of 30 cubits did compass it round about.

The same description is repeated almost verbatim in Chapter 4 of the second book of Chronicles—apparently someone was copying without checking the arithmetic. If the "sea" (probably a circular fountain or basin) is indeed a circle with a diameter of 10 cubits, then the circumference must be $\pi \times 10$ cubits, or approximately 31.4 cubits. If the biblical cubit is approximately 45 centimeters (or roughly 18 inches), the biblical calculation involves a discrepancy of more than 60 cm.

The number π is, after all, not exactly 3 (even if the legislature in the state of Indiana voted it so), and in practice "π is approximately 3" rarely suffices. But how precise do we really need to be, if a thumb and a pi is not enough? In school we learned that π is approximately 3.14, or that it is close to the fraction $\frac{22}{7}$; in both cases the error is about one and a half parts per thousand. For the biblical fountain, that would be an error of a few millimeters. Is that precise enough?

In *practice* that is a question of how it is to be applied. For a brass fountain, an unplanned gap of several millimeters would not have been desirable even in Solomon's time. It is little wonder, then, that modern technology, which ranges from atomic dimensions (in semiconductor devices) to astronomical distances (for satellite communications), requires a few more digits after the decimal point. On the other hand, for ordinary household computations, the ten digits that every hand-held calculator provides will probably suffice.

For *theory*, however, this is certainly not enough. We have known since the eighteenth century that π is not rational, that is, cannot be expressed as a ratio of natural numbers—Johann Heinrich Lambert proved this in 1761. Therefore, expressions such as 22/7 or 355/113 or 3.1415926535 (also a fraction: 31, 415, 926, 535/10,000, 000, 000) are only approximations. The decimal expansion

3.141 592 653 589 793 238 462 643 383 279 504 884 197 169 399 375 105 820

never terminates and never begins to repeat itself; it never becomes "periodic," no matter how long we wait and how hard we look.

The number π has fascinated people since the beginnings of civilization—and still does. It is not just the problem of computing the number as precisely as possible that makes it interesting. For example, an Italian named Pietro Mengoli asked around 1644 if one takes all the squares of the integers, computes the reciprocals, and adds them up, does one get a finite answer? In other words, he asked if the sum

$$1 + \frac{1}{4} + \frac{1}{9} + \frac{1}{16} + \frac{1}{25} + \frac{1}{36} + \frac{1}{49} + \frac{1}{64} + \frac{1}{81} + \frac{1}{100} + \frac{1}{121} + \frac{1}{144} + \frac{1}{169} + \frac{1}{196} + \frac{1}{225}$$

is finite. The question was later referred to as the Basel Problem because several mathematicians from Basel gnawed at the problem, but without success. Among these were several members of the Bernoulli family, who resided in that Swiss city. Finally in 1735, Leonhard Euler (1707–1783; also from Basel), who had been the successor of Daniel Bernoulli as mathematics professor at St. Petersburg since 1733, was able to present the answer to the problem after ten years of work. The infinite series has a finite sum: it is $\pi^2/6$. If one asks mathematicians for the most beautiful formula in which π appears, one will get a number of different answers, but Euler's solution of the Basel Problem will often be named.

One might suppose that such a fundamental constant as π would have been researched comprehensively and is now completely understood, but that is not at all the case, and will never be the case. Mathematics is not a finite, closed-off discipline, and while research provides answers to some questions, these answers themselves raise new questions—sometimes simple, but sometimes complex, and occasionally unsolvable. The following question, for example, seems harmless, but has not yet been answered: in the decimal expansion for π:

3.141 592 653 589 793 238 462 643 383 279 504 884 197 169 399 375 105 820 974 944

do all the digits (and pairs of digits, triplets, and longer sequences of digits of a certain length) appear equally often? And is the sequence of digits statistically indistinguishable from a truly random sequence? Mathematicians believe this, and indeed, nothing in an analysis of the first billion digits suggests otherwise. There is no evidence, for example, that 8 appears more often than it should (one tenth of the time), or that the triplet 123 appears less than other triplets. Statistically, the sequence 999999 should appear somewhere in the billion digits, and indeed it does, the first time starting at the 726th digit in the sequence, called the Feynman Point. (The name comes from the fact that the American physicist Richard Feynman said—in his memoir, *Surely You Are Joking, Mr. Feynman*—that he wanted to memorize the digits of π to the point that he could stop the recitation with "nine, nine, nine, nine, nine, nine.") It seems possible to memorize 726 digits, even for ordinary mortals, I should think, especially with a little memory training. I have never seriously tried it myself.

Daniel Tammet, the young, autistic British memoirist we've already met, actually finds it easy to memorize such long strings of numbers. Since March 14, 2004, he holds the European record for memorizing digits of π, having recited, without a single error, 22,514 digits. He says he sees the digits as pictures, colors, and shapes, and the sequence as a fascinating landscape. The Feynman Point is, for him, a beautiful sight, a deep, thick border of dark blue light. Indeed, he finds π particularly beautiful and unique. Like the Mona Lisa or a symphony by Mozart, π is, of itself, beautiful enough to love.

Actually, π probably has everything a pop star needs: everyone is familiar with it, no one knows it well, and an air of mystery surrounds it. In some places in the United States it is celebrated orgiastically on March 14—written as 3.14 instead of 3/14 for the occasion and renamed π-day. But even more, I like the secret admirer, perhaps the π-fan to whom Kate Bush sings in her album *aerial*:

> Sweet, and gentle, and sensitive man,
> with an obsessive nature and deep fascination for
>   numbers
> and a complete infatuation with the calculation of π

and in whose honor the singer then follows these lines by a recitation of the first 120 digits of π, unfortunately with a few missing and erroneous digits. But she's allowed to do that.

$\sqrt{-1}$—Victim of a Character Assassination

What haven't we had to put up with! After all, the natural numbers—1, 2, 3, 4, …—give us a certain and clear impression of what it means to be a number. The negative numbers take a little getting used to, but we've managed to accommodate ourselves to them (even if we still find them problematic when they appear in our bank accounts). We have managed to develop an idea of two thirds, and perhaps even the taxes of 8.75/100 or 19/100. Computation with fractions is something we can do after a little practice and can mostly be avoided in the ordinary course of things, especially since all the cash registers now do it automatically. We are used to dealing with decimal expansions almost hourly, since prices are listed as decimal expansions to two digits (dollars and cents)—except for gasoline, where the prices are given to a tenth of a cent, and the occasional supermarket items that involve a half a cent, as when lemons are two for 95 cents, so each lemon is $0.475.

What happens if we are asked to think of an "arbitrary" number? Do we then think of 3, or 42, or 1.99, or $\sqrt{2} = 1.4142...$, or $\pi = 3.14159...$? Do people develop their own ideas of a number in this way? Perhaps sorting them out on a line, in order, with zero in the middle, positive numbers to the right and big positive numbers to the far right. This does seem to be the case. Modern medicine confirms it with, among other results, magnetic resonance imaging scans that show that the same sections of the brain are activated when one is dealing with large numbers as when one is instructed to "look right." The number line, then, seems to be a pretty complete notion of the concept of "number."

If, however, numbers are "the things that mathematicians use to compute with," then there are considerably more numbers, and a much greater variety of numbers, than we see in our ordinary computations. Among these are the "complex" numbers, which are numbers like $4 + 2\sqrt{-1}$. One can actually compute with these numbers consistently and systematically. They are also indispensable for modern mathematics, for example, in the solutions of equations (a high point of which is the *Fundamental Theorem of Algebra*, which was proved by Gauss in his PhD thesis in 1799); they appear in the analysis of oscillations and waves, or tones and overtones (what mathematicians call Fourier analysis); and they are used in higher algebra and number theory. Complex numbers are also essential in many areas of physics (such as quantum theory) and in engineering.

They are, however, the victims of character assassination. In 1637, the French mathematician and philosopher René Descartes (1596–1650) called them "imaginary," and the term of opprobrium stuck. An aura of unreality and difficulty has beset these numbers ever since. Nowadays we call them "complex" numbers, which is hardly better, and we still refer to $i = \sqrt{-1}$ as the "imaginary unit." Would that Descartes had called them "complete numbers" or some other tangible concept.

The general opinion of these numbers has never, I believe, completely recovered from Descartes's slander. Because of their name, they are commonly deemed not suitable for instruction in the schools but reserved for adults. This is too bad. For example, the analysis of waves and oscillations is a marvelous area of mathematics (and also of physics and music), for it deals with the science of harmony. Unfortunately it was not until 1799 that it was recognized (by Caspar Wessel) that the complex numbers could be displayed elegantly as points in a plane. Today this is called the Gaussian number plane; it is an infinite plane divided in two by a line, the axis of "real" numbers, like a straight railroad track dividing a landscape. A traveler along this line who looks only at the tracks, at the line of real numbers, and does not look to the left or right will miss a great deal of the landscape, all of it very interesting. Or are you one of those travelers who is satisfied that there is entirely enough to see in the train itself?

$\aleph_0$—The End of the Number Line?

The infinite has been a matter of interest to philosophers since antiquity. On the other hand, mathematicians in general thought it was stuff and nonsense, and the first attempts to consider the subject from a mathematical and rigorous point of view were dismissed rather forcefully. I suspect that even here we let the aura surrounding the names influence our opinions of the concepts.

The infinite does not exist, or if it does, it is a hopelessly abstract concept. Infinite sets, however, certainly do exist and, in fact, do not present any conceptual problems. There are, for example, infinitely many natural numbers:

$$1, 2, 3, 4, 5, 6, 7, 8, 9, 10, 11, 12, 13, 14, 15, 16, 17, 18, 19, 20, 21, 2$$

which we can consider collectively as a set: the set of all natural numbers, a set with infinitely many elements.

Georg Cantor, born in 1845 in St. Petersburg to a stockbroker, married and the father of six children, is considered, as I have mentioned, to be the father of set theory. The subject is not at all abstract stuff and nonsense. Quite the contrary, Cantor showed us how to make the subject concrete. When are two sets equally large? Clearly, when we can put all the elements of one set into a precise one-to-one correspondence with the elements of the other. For infinite sets, this definition works equally well, but it leads to some marvelous places, among them Hilbert's Hotel.

This hotel was invented by David Hilbert (1862–1943), who used it to illustrate Cantor's ideas. Imagine a hotel that has not just a lot of rooms, but an infinite number; the rooms are numbered sequentially:

$$1, 2, 3, 4, 5, 6, 7, 8, 9, 10, 11, 12, 13, 14, 15, 16, 17, 18, 19, 20, 21,$$

Such a hotel cannot be constructed in practice, but one can imagine one of these incredibly large hotels in some European metropolis with corridors that just go on and on; and if one squints a little, they seem to go on forever. (A mathematician would call the number of rooms in this hotel a "countable infinity" and denote it by $\aleph_0$.)

Now let us suppose that every room in Hilbert's Hotel is occupied. A guest arrives. What to do? The hotel manager can accommodate them without a problem (albeit with a little effort). The guest in Room 1 is simply moved to Room 2, the guest in Room 2 is moved to Room 3, and so forth. The new guest is then moved into Room 1. No problem. But what if a large group of guests arrive, an *infinitely* large group? If the hotel manager knows his math, he realizes he can simply move the guest from Room 1 to Room 2, the guest from Room 2 to Room 4, the guest from Room 3 to Room 6, and so forth; this will free up all rooms with an odd number on the door. And what does our (steel-nerved) hotel manager do with an *infinite* number of such groups? He can find a solution. And so can you.

One can use the hotel manager's schemes in other directions as well. For example, if the hotel is full, but corporate headquarters insists the rooms must be used more effectively, the manager can ask two guests to double up in Room 2, three guests to occupy Room 3, and so forth. This also has a concrete solution, and it ensures that the bellhops will be kept very busy in Hilbert's Hotel.

Could we use decimal expansions to number the rooms in the hotel? That is, could there be a room 0.42 as well as a room 0.33333…? As long as we consider only finite or repeating decimal expansions, we have no problems. But what about a room labeled $\sqrt{5}$ or π? An elegant proof devised by Cantor, his "diagonal argument," shows that we cannot accommodate all the real numbers in the rooms of Hilbert's Hotel. In other words: the real numbers cannot be counted (whereas the rational numbers—fractions—can all be counted). This is a surprising result, and it changes one's concept of "infinity," for it is now clear that there are infinite sets of different sizes.

Note that I am referring to infinite sets, and not infinity itself. I have no concept of "infinity"; it is purely abstract, and one cannot imagine something that is purely abstract. Or maybe one can?

> And suddenly Arthur had a fairly clear idea of what infinity looked like.
> It wasn't infinity in fact. Infinity itself looks flat and uninteresting. Looking up into the night sky is looking into infinity—distance is incomprehensible and therefore meaningless. The chamber into which the aircar emerged was anything but infinite, it was just very very very big, so big that it gave the impression of infinity much better than infinity itself.

Thus the expert, Arthur Dent, who experienced the infinite reaches of the universe with *The Hitchhiker's Guide to the Galaxy.*

Chapter 2

The Never-Ending Story of Prime Numbers

What sequence of numbers begins with 2, 3, 5, 7? For us, the first thing that comes to mind is the prime numbers. And that is a correct answer, but certainly not the only correct answer. If you consult the *On-Line Encyclopedia of Integer Sequences* (oeis.org), which was compiled by the mathematician Neil J. A. Sloane at the legendary Bell Labs, you will find 1984 sequences that start this way (as of March 2013).

The first of the 1984 sequences is indeed the prime numbers:

2, 3, 5, 7, 11, 13, 17, 19, 23, 29, 31, 37, 41, 43, 47, 53, 59, 61, 67, 71, 73, 79, 83, 89, 97

The second of the sequences is the number of partitions of the integers:

2, 3, 5, 7, 11, 15, 22, 30, 42, 56, 77, 101, 135, 176, 231, 297, 385, 490, 627, 792, 1,002

That is, 2 can be expressed as two possible sums, just by itself, or as 1+1; 3 can be written in three ways: 3, 2+1, and

1+1+1; 4 has five partitions: 4, 3+1, 2+2, 2+1+1, and 1+1+1+1; 5 has seven partitions (whose enumeration I leave to you), and so forth. This sequence of partitions is interesting, and not only because the number 42 appears in it. It has been studied extensively in the last 200 years—but in that respect it lags far behind the primes, which have been studied for thousands of years and by nearly *all* the giants of mathematics. After the discovery of the first four prime numbers—2, 3, 5, and 7—in deep prehistoric times, and the next four—11, 13, 17, and 19—which appear as markings on the Ishango bone from 20,000 years ago, there soon arose the question, how many of these prime numbers are there altogether?

Euclid Is Still Right

In Euclid's *Elements* there is a proof that there must be an *infinite* number of primes, that is, that the sequence

2, 3, 5, 7, 11, 13, 17, 19, 23, 29, 31, 37, 41, 43, 47, 53, 59, 61, 67, 71, 73, 79, 83, 89, 97, 1(

simply does not end. The proof is simple; it is even taught in schools. Even more: it is the very *model* of a proof.

Euclid's *Elements* is not only the most successful textbook in the history of mathematics but also the most important. About Euclid himself, however, we know almost nothing other than that he was active around 300 BCE, possibly studied at Plato's Academy, and is supposed to have written the greater part of the *Elements* himself. For many hundreds of years, this work has been used to teach and to learn how to construct proofs, that is, how to draw valid consequences from clear fundamental assumptions (axioms) and thereby reach sound conclusions (theorems). After studying Euclid, one knows what a valid mathematical proof is. Just as Germans can say that a computation is correct "according to Adam Ries," we can also say that a proof is valid "according to Euclid."

A few years ago, Noga Alon, a mathematician at Tel Aviv University, was interviewed on an Israeli radio station about mathematics. He explained to the interviewer that Euclid had already shown 2300 years ago that there is an infinite number of primes. "And," asked the moderator, "is this still true?" We can assure the moderator that it is indeed still true. That's the beauty of mathematics. If a computation is valid "according to Adam Ries," it remains so. If a proof is valid "according to Euclid," it is valid forever. In this sense, mathematics is reliable and stable and offers eternal truths. There is no "Adenauer learning effect" through which yesterday's comments become today's uninteresting drivel. (German Chancellor Konrad Adenauer famously responded to an inconsistency in his current comments with remarks he had made before, "Why should I be interested in the claptrap I said yesterday?" One should, however, add that Adenauer did not really apply that comment to his own statements. While he always insisted in his speeches on his right to become smarter with time—"a fundamental human right"—and on the right to learn from his political opponents, he never referred to the opinions he rejected as "claptrap." At least, so we are told by Christoph Drösser in one of his "check the myths" ("Stimmt's") columns in the weekly newspaper *Die Zeit*. In Euclid's case, the question never arises.)

How Many Prime Numbers Are There?

Infinitely many, of course, as we know from Euclid. But is that enough? Every good answer creates new questions. Like, how many prime numbers are there with three digits, that is, between 100 and 1000, or with four digits? Euclid's proof has nothing to say about this.

One way to answer that question is by simple enumeration. This works at least for the questions above. We can count 4 primes with a single digit, 21 primes with two digits, 143 with three digits (that is, between 100 and 1000), 1161 with four digits, 8363 with five digits, and so forth. And so forth? So far the record, set by Tomás Oliveira e Silva in 2007, is the number of 23-digit primes, exactly 1,723,853,104,917,488,062,633. But how many there are with 30 or 40 digits, no one can say precisely.

When we can't know something precisely (or maybe don't really want to know it precisely), we can make an estimate—perhaps an estimate that is as precise as possible, but still only an estimate. Let us use a notation that has become common in number theory since the time of Euler and denote by $\pi(1000)$, the number of primes smaller than 1000. Then $\pi(1000) = 168$ is the number of primes with at most three digits, $\pi(100) = 25$ is the number of primes with at most two digits, and $\pi(1000) - \pi(100) = 168 - 25 = 143$ is the number of primes with exactly three digits. So, about how large is $\pi(10^n)$, the number of primes with at most n digits?

When he was only fifteen years old, in 1793, Karl Friedrich Gauss wrote a conjecture into his diary about the fraction of numbers that are prime, namely, that of all numbers with n digits, approximately $\frac{1}{n} \ln 10$ are prime; here $\ln 10 = 2.302505\dots$ is the "natural" logarithm of 10. Gauss based his conjecture on an extensive list of primes that he compiled himself (computing them by hand, of course). Independently in Paris in 1798, Adrien-Marie Legendre (aged forty-six) arrived at the same conjecture. Thus, for example, of all the numbers with at most ten digits, approximately $\frac{1}{23}$ are prime, while of the numbers with at most 100 digits, only one in 230, or about 0.4%, are prime. Another way to interpret the conjecture is that an arbitrary number of at most n digits has a probability of 1 in $2.302585 \times n$ of being a prime number.

At the time that Gauss and Legendre put forth their conjecture, confirming it lay far beyond the reach of the available mathematical methods. A hundred years later, developments in analysis—and in particular, developments in the theory of complex functions and, building on that, analytic number theory—changed the situation. In 1896 Gauss's conjecture was proved independently by two mathematicians, Jacques Hadamard in France and Charles Jean de la Vallée Poussin in Belgium. No longer a conjecture, it has since been called the Great Prime Number Theorem.

The Great Prime Number Theorem is beautiful and important, and it is certainly a masterly achievement of nineteenth-century mathematics. But for a serious mathematician, it is no reason to set the question aside: We would like to know how accurate the approximation is. And since it is only an approximation, are there better approximations? The answer to that is presumably yes: the Riemann Hypothesis provides a considerably more precise estimate, the so-called logarithmic integral function, and a guarantee that the error—that is, the difference between $\pi(n)$ and the estimate—is a number with at most half as many digits as the estimate. Unfortunately there is a catch: the Riemann Hypothesis, which was put forth by the brilliant young mathematician Bernhard Riemann in Göttingen in 1859, has not yet been proven. It has for a long time now been one of *the* unsolved problems of mathematics, listed, for example, in 1900 among the twenty-three most important unsolved problems in mathematics, and listed again 100 years later among the seven Millennium Problems, each of which the Clay Foundation offered a prize of a million dollars for the solution. The jackpot is still available. We are waiting for a breakthrough.

Fermat Made a Mistake

When you square the number 2 (that is, multiply it by itself) you get 4; when you square 4, you get 16; squaring that gives 256; and if you keep doing that you get a sequence

$$2, 4, 16, 256, 65{,}536, 4{,}294{,}967{,}296, \ldots$$

and your standard pocket calculator will probably not be able to fully display the next number in the sequence because it has twenty digits—each number in the sequence has twice as many, or at most one fewer than that, digits as its predecessor. More interesting than this sequence of rapidly increasing powers of 2, which we can write as 2^{2^n}, are the numbers obtained by adding 1 to each of them:

$$3, 5, 17, 257, 65{,}537, 4{,}294{,}967{,}297, \ldots$$

which can be described by the formula $2^{2^n} + 1$. One can easily see that each of these numbers ends with a 7 (except for the first two). And perhaps you also notice something else interesting about them. Do you? Do you think they look like prime numbers? Indeed, Pierre de Fermat claimed that numbers of the form $2^{2^n} + 1$ are all primes, and for the first five numbers, at least, he was able to show by computation that they are indeed primes.

Pierre de Fermat (c. 1607–1665) was trained in law and worked as a lawyer and judge, but we also have to thank him for several important contributions to mathematics—and one of the greatest puzzles, the famous "Last Theorem." We also owe one of the most interesting errors in mathematics to him, namely that the numbers 3, 5, 17, 257, 65,537, … are all prime numbers. That this is not the case was first noticed by Leonhard Euler, who investigated the Fermat primes $2^{2^n} + 1$ in St. Petersburg at the age of twenty-five. He found that 4,294,967,297 is divisible by 641 and is thus not a prime number. This is not something one can immediately recognize. On the other hand, Eric T. Bell, in his classic *Men of Mathematics*, tells about a lightning calculator named Zerah Colburn who could, after only a short time, say that the number 4,294,967,297 is divisible by 641. How he arrived at that result, however, the numbers juggler did not divulge.

Even if we can't match Mr. Colburn in the speed with which he recognized the divisibility, do we at least trust ourselves to understand what it means? And to know how one could confirm it?

In the meantime, it has been found that not only is this one Fermat number 4,294,967,297 not a prime, but that at least the next twenty-seven Fermat numbers are also not primes. There are also no known Fermat primes beyond that. One is thus led to suppose that there are no Fermat primes beyond the first five, but no one has been able to prove that.

Why is this interesting? Because the Fermat primes are connected with one of the classic puzzles of geometry: which regular polygons can be constructed only with ruler and compass? The answer is due to Karl Friedrich Gauss and is one of his early masterpieces (the corresponding bed story appears later in this book): one can construct a regular polygon having an odd number n of sides with ruler and compass only when n is a product of different Fermat primes. So one can construct a regular 17-sided polygon with ruler and compass, but not a regular 7- or 9-sided polygon. And also not a regular 4,294,967,296-sided polygon.

Mathematics majors learn Gauss's theorem about constructing regular polygons in their first serious algebra class. The proof is not complicated using contemporary methods, but the answer is still not complete because we don't know whether there are further Fermat primes.

The "Mozart of Mathematics" Makes Use of an Error

Mathematics is the science of patterns. That is why the "chaotic" sequence of prime numbers

2, 3, 5, 7, 11, 13, 17, 19, 23, 29, 31, 37, 41, 43, 47, 53, 59, 61, 67, 71, 73, 79, 83,

is such a great challenge. There is much to seek and much to discover in that "chaos."

Among the first things one notices are the twin primes, pairs of primes that are separated by just one number. At the beginning of the sequence we see many twins: 3 and 5, 5 and 7, 11 and 13, 17 and 19, 29 and 31, and so forth. And so forth? Does that mean that there are infinitely many twin primes? How frequent are they? These are unanswered questions. We believe that there are infinitely many twin primes, and there are even fairly precise estimates for their frequency. But none of that is proven. Can you find a large pair of twin primes? By hand? With a pocket calculator? I don't want to discourage anyone from trying, but currently the largest twin primes, found in December 2011, have 200,700 digits each. By the time you read this book there may be even larger examples—just Google "twin primes."

Sequences of "constant gap length" are another interesting kind of pattern. These are sequences of primes in which each is separated from the others by the same amount. Mathematicians call such sequences "arithmetic sequences." The primes 3, 5, 7, for example, are such a sequence with a length of three (and a gap between the primes of 2); the primes 5, 11, 17, 23, 29 are a sequence of length five with a gap of 6. Can you find a longer sequence? No? These are not easy to find "manually."

Every arithmetic sequence of primes must terminate—and usually they terminate quite soon. But even if there are no unending arithmetic sequences of primes, there could be some very long ones. The longest one known at present comprises an astonishing twenty-five prime numbers. Can there be still longer sequences—arbitrarily long sequences? Indeed there are. This was proven in 2004 by the British mathematician Ben Green and the Australian Terence Tao. Green was twenty-seven years old at the time and Tao was one and a half years older. While Green kept himself in the background, Tao was not able to avoid public notice. He received the Fields Medal, mathematics' highest honor, at the 2006 International Congress of Mathematics for his achievements, including the result on prime-number sequences. One of Tao's colleagues at UCLA was quoted as saying that mathematics flows from his pen like music from Mozart's. One can be at most forty years old to receive a Fields Medal; this was not a barrier for Terence Tao, the son of Chinese immigrants, who won a Gold Medal at the International Mathematics Olympiad at age twelve, received his PhD from Princeton at age twenty-one, accepted a professorship at UCLA when he was twenty-four, and was only thirty-one when he was honored with the Fields Medal. In addition he is a nice person, humorous, witty, and gifted in languages. Not an eccentric genius. "He has a lower neurotic index than most of us," his thesis advisor told the magazine *Spiegel.*

Some of what Tao achieves in mathematics is ingeniously simple. The work with the prime numbers, however, was ingeniously complicated. And it was based on an error: in March 2003, Don Goldston of San Jose and Yalçın Yıldırım from Istanbul presented a result at a conference at the Mathematical Research Institute Oberwolfach in the Black Forest (a place that you will get to know better later in this book) according to which there are arbitrarily large prime number pairs with "relatively small separations." They could not prove that the separation was always 2 (that is, that there is an infinite number of twin primes), but the result was heralded as a breakthrough.

When results are particularly interesting, colleagues will look at them particularly carefully. That happened here as well, and they found a subtle error in one approximation that immediately brought the whole proof into question: if a single link in a chain fails, the whole chain cannot hold.

Remarkably, Green and Tao were able to rescue a decisive idea from the "broken" proof of Goldston and Yıldırım and use it as a cornerstone for the structure of their proof—a structure that used other, highly developed resources from many diverse areas of mathematics whose names are probably familiar only to experts: analytic number theory, harmonic analysis, combinatorics, ergodic theory. In short, the Green–Tao proof is a masterpiece. (I will say no more about it. Because if you ask me if I really understand it, then I would have to answer, I've not even tried, for it would surely take months of study of areas of mathematics with which I am not at all familiar—and in this case, that includes all the fields I mentioned except for combinatorics. Mathematics is a huge and multifaceted field, and any one mathematician who is not Tao can have an overview and mastery of only a small part of it. But maybe no one will ask me.)

One can supplement the story with a happy ending also for Goldston and Yıldırım. After several attempts they were able, together with the Hungarian Janos Pintz from Budapest, to fix their proof about small gaps between prime numbers; the proof was completed in February 2005, was much simpler than the previous attempt, and withstood all critiques. The question of twin primes (that is, gaps of size 2) remains unanswered.

Another Search for Errors

Since Euclid, we have known that the number of primes is infinite. And more precisely, or more ominously, it is not difficult to show that the sum of the reciprocals of all primes

$$\frac{1}{2}+\frac{1}{3}+\frac{1}{5}+\frac{1}{7}+\frac{1}{11}+\frac{1}{13}+\frac{1}{17}+\frac{1}{19}+\frac{1}{23}+\frac{1}{29}+\frac{1}{31}+\frac{1}{37}+\frac{1}{41}+\frac{1}{43}+\frac{1}{53}+\frac{1}{59}+\frac{1}{61}+$$

does not converge: that is, it exceeds all bounds if one adds enough terms. Euler was able to prove this.

On the other hand, for twin primes it is the opposite. There is as yet no proof that there is an infinite number of them. Nonetheless the sum

$$\frac{1}{3}+\frac{1}{5}+\frac{1}{7}+\frac{1}{11}+\frac{1}{13}+\frac{1}{17}+\frac{1}{19}+\frac{1}{29}+\frac{1}{31}+\frac{1}{41}+\frac{1}{43}+\frac{1}{59}+\frac{1}{61}+\frac{1}{71}+\frac{1}{73}+\frac{1}{101}+\frac{1}{103}+$$

is finite. Even if, as one currently expects, there is an infinite number of terms to be added up, they all add up to a finite number. The value of that sum is called Brun's Constant, B, after the Norwegian Viggo Brun (1885–1978), who proved this in 1919. Adding the first eight terms in the sum gives the approximate answer of 1.155, but this is far from the best we can do. A much more precise estimate, $B = 1.90216054$, was provided by the Australian Richard Brent, who used tables of twin primes with up to twelve digits. One Thomas R. Nicely, working at Lynchburg College in Virginia, was able to go yet further with the computations and published in 1995 the result that $B = 1.90216058$, a little larger than Brent's result in the eighth decimal place.

Which is nice, and an interesting result. But the story is even more interesting for an entirely different reason. In the course of his computations, Nicely found several errors in his computer output, errors that he could detect only because he performed all of his computations on two different machines built with different processors and running different computer codes. The most serious error turned out to be a bug in the brand new Pentium chip that Intel had introduced in 1993. This now-notorious "Pentium bug" led to rare errors in multiplication, so that, for example

$$\frac{1}{824,633,702,441} + \frac{1}{824,633,702,443}$$

was not correct, as promised—and needed for Nicely's computations—to nineteen decimal places, but only to nine. That is a major difference if one is adding many numbers and needs to avoid accumulating too many errors (rounding to nineteen decimals already introduces some errors, for example).

The Pentium bug turned into a middling catastrophe for Intel. Not only was it a public relations debacle, the company wound up having to redesign the chip and replace millions of them in computers that had already been sold, and they had to admit that they had found the problem in the course of testing the chip and had kept silent about it, hoping no one would notice. Mistake: there will always be mathematicians who would like to know something very precisely, and therefore have to compute with great precision. The precision needed by prime-number researchers occasionally exceeds even the precision needed by computer-chip designers.

A Well-Insured Million

We owe one of the most famous unsolved problems of mathematics, Goldbach's Conjecture, to Christian Goldbach (1690–1764). Goldbach was, like Leonhard Euler, active for many years at the Academy of Sciences in St. Petersburg. In 1741 Euler moved to Berlin, following Frederick the Great's invitation to join the Prussian Academy of Sciences. One year later, in one of his letters to Euler, Goldbach asked if any even number (other than 2) could be written as a sum of two *numeris primis*—that is, prime numbers. We can try a few samples and see how it works: $4 = 2 + 2$, $6 = 3 + 3$, $8 = 3 + 5$, $10 = 5 + 5$ or $10 = 3 + 7$, $12 = 5 + 7$, $14 = 3 + 11$, or $14 = 7 + 7$, and so forth. And so forth? That is precisely Goldbach's question.

These days, one can, of course, test the conjecture quite extensively with a computer. As of 2007 it has been found correct for all numbers of at most eighteen decimal places. In addition, plausibility arguments of analytic and probabilistic number theory indicate that the conjecture should be correct. But it is not proven.

Asking questions is easy. Answering them is hard. Goldbach's Conjecture is extremely easy to formulate, but it is apparently really hard to prove. Not only a large number of mathematicians but even heroes of novels have broken their teeth on it. Among the latter is Professor Petros Papachristos, the uncle of the narrator and tragic hero of the novel *Uncle Petros and Goldbach's Conjecture* by Apostolos Doxiadis. Uncle Petros does not let us know by the end of the novel whether he has a proof or can find one. When the novel was published in English in 2000, its publishers, Faber and Faber in the United Kingdom and Bloomsbury in the United States, jointly offered a reward of $1 million for a proof of the Goldbach Conjecture. One can assume that this was not a serious attempt to further research but rather a (clever) marketing ploy. That is confirmed by the deadline set by the publishers for the prize: the proof had to be completed, reviewed, and published within two years. One can go on the assumption that there is no simple proof of Goldbach's Conjecture, and that a proof would not be found in a trice. And even if someone made a decisive breakthrough tomorrow or the day after (having perhaps worked on the problem for years), the work of completing it, writing it down, having colleagues check it, submitting it to a technical journal, waiting for the anonymous peer review, and finally publication would hardly be possible in two years. Nonetheless, the publishers were not entirely certain of their prospects and, one learned, took out an insurance policy, which is the prudent thing for a business to do. I do not know what premiums the insurance company required for the policy, but it would be interesting to know because one could then see how (un)likely the insurance company's experts in risk analysis thought the risk of a payout to be.

Chapter 3

The Mathematical Perspective

Numbers determine our fate. I don't mean that with any ominous undertones, I am merely referring to the numbers in our bank accounts, in our tax forms, in the statistics we read about, and in the bills we pay.

But if numbers are really so important, then it would be worthwhile to examine them carefully: are they plausible, believable, correct? Some 82.49% of Americans deal much too uncritically with numbers; they fall for the specious authority and exactitude of numbers—even when these are sometimes totally invented, such as the percentage at the beginning of this sentence. A healthy dose of numerical skepticism never hurt anyone. Occasionally one can see immediately that a number is a lie. But in general we let ourselves be dazzled and dispense with the necessary skepticism when presented with particularly impressive numbers. We should always check on the order of magnitude, review the level of precision, and, when necessary, see if we can reproduce a computation. We should mistrust formulas until we know exactly what they say. We should approach numbers not with credulity but with a cheerful, persistent search for errors. Common sense. And a mathematical perspective.

Estimates

We all know what a number is, but what is a numerical estimate? If the organizers of a demonstration know very well that they attracted 100,000 participants, but the police will admit to counting only 15,000, who is correct? Which number do you believe? Does that depend on whether you attended the demonstration? Would you find it more plausible if the organizers claimed 102,296 participants? Or if the police counted only 14,956? According to what criteria are these numbers—estimates, in fact, since no one can actually count all the individuals who attend a demonstration—arrived at?

For an indoor concert, an attendance figure is fairly easy to determine: one knows how many tickets were sold and how many people fit into the hall. Even journalists attending the performance can practice their estimation techniques: "About thirty rows, each with about thirty seats, but in general about half occupied, so that makes roughly 500 attendees." The "roughly" comes about because of the multiplication of three numbers, none of which is precise. If each is off by 10% from the proper value (which, after all, can easily happen), then the result can be too small or too large by more than 30%. However, in most cases one of the estimates will be too small, another will be too large, and the net result will be about right.

For outdoor events, open-air concerts, or political rallies or demonstrations, such estimates are much harder to make. There are two reasons for this: first, the numbers are more difficult to estimate, and, second, there are vested interests (particularly for political events) in reporting impressively small or large attendance numbers. How did that go, for example, when the then-candidate Barack Obama spoke at the Victory

Column in Berlin in 2008? When the Austrian Press Agency (APA) reported, "Barack Obama inspired more than 200,000 listeners at Berlin's Victory Column," we should examine this with our magnifying glass and focus on the phrase "inspired more than 200,000 listeners." When the same APA release said, "In front of more than 200,000 cheering supporters, he challenged Americans and Europeans to put aside their conflicts and to struggle together against global problems such as climate change and terrorism," we can only agree. On the other hand, one could also turn to more skeptical reports, such as the one in the *Frankfurter Allgemeine Zeitung*, which reported, "Before ten thousand visitors, the Democratic senator called out, 'People of the world, look at Berlin!' With these words he recalled the historic speech of the former mayor of Berlin Ernst Reuter, who spoke to more than 300,000 people in front of the Reichstag during the Berlin Blockade of 1948–49."

So, were there 200,000 or only 10,000 people at the Obama rally in 2008? (Never mind that the audience of 10,000 looks even more puny when compared with the 300,000 from the past. That's another clever technique of political rhetoric.) I don't know which the answer might be, even though I was myself in that crowd. The police, who were of course there to watch over the event, could estimate the numbers fairly easily. The square directly around the Victory Column, which was closed off with barriers and access controls, can indeed hold only around 10,000 people. But then there were also masses of people along the Strasse des 17 Juni; this street, which runs from the Brandenburg Gate for several kilometers through the Tiergarten park, is a wide boulevard, with several lanes, side roads, bicycle paths, and walkways making it some 35 m wide. People stood here, tightly packed together—even 500 m from the Victory Column the street was still crowded. The total area covered densely with people was thus probably 20,000 square meters (roughly 200,000 square feet); if we figure an average of four people per square meter, then the total number of people at the rally would be something like 100,000, but probably not 200,000. Of this number, most could probably not see the stage but only, at best, one of the large LCD screens, or possibly just the wide shoulders or the large hairdo of the person in front of them.

For this estimate, I also had to multiply three numbers: the 35 meters is reasonably precise; the 500 meters is also a reasonably good estimate, since the crowd stretched halfway to the Brandenburg Gate, which is about 1000 m from the Victory Column. Less certain is the number of people per square meter. If you don't trust my rough guess, you can find a photo of the rally (an aerial photo is eminently suitable—and for large assemblages the police and press have these available), mark on it several "typical" squares of, say, 10 m by 10 m, and then count the people in each and use the average for your estimate of the number in the crowd. This method provides quite reliable estimates, which can then be further "massaged" before publication.

The same method of dividing a large area into representative squares is also the one used to count red and white blood cells under a microscope—and spermatozoa. These counts are then reported to the patient. Or not.

Furthermore: the *real* number of participants in a demonstration or rally does not, in fact, exist. Think of the presidential inaugurations or large parades such as St. Patrick's Day parades or Gay Liberation marches in the large cities. Who is a participant, who is an onlooker? And what about the participants who get sore feet and leave the crowd, and the others who were stuck in traffic and arrive halfway through? Do we count them both, since both participated? Or are we interested only in the number who were there at any one time or simultaneously?

Random Numbers

Chance rules the world—this is true, but not the whole truth. Many of the numbers that surround us are random, or at least look random. The stock market index values, for example, result from the addition of many small and large transactions that are in the end not computable or predictable and thus look random.

However, it is not really easy to generate random numbers, numbers that *look* as if they were generated randomly. You can try it yourself: write down a sequence of hundred numbers (single digits, between 0 and 9), each number completely independent of the ones you wrote down previously and the ones you will write subsequently. When you've done that, look it over. Is the sequence truly random?

Let's look at three sets of digits:

First

141 592 653 589 793 238 462 643 383 279 502 884 197 169 399 375 105 820 974 944 592 307 816 406 286 208 998 628 034 825 342 117 067 9

Second

022 424 170 465 800 187 484 459 427 552 056 869 842 622 684 442 412 286 285 840 996 417 214 180 386 408 307 450 694 607 189 265 773 0

Third

482 750 184 728 948 288 205 729 472 991 045 577 749 930 275 922 784 958 583 011 038 572 038 491 102 347 857 309 847 583 938 851 200 3

If these are truly sequences generated by chance, they must satisfy several criteria. For a *real* random sequence, all ten numbers should appear with pretty nearly equal frequency—but not *precisely* equal in any finite sequence, just pretty nearly. If the distribution were perfectly uniform, the sequence would not be random. A random sequence should have, on average, two equal digits in any group of ten digits. Any group of ten digits should also contain a pair of sequential numbers (that is a pair like 01, 12, 23, ... 78, 89, 10). There are several similar statistical tests that should not show any unexpected groupings or patterns of digits.

Let us then look at the three sequences above. Which of them looks convincingly random? Which does not? At first glance, they all look as if they were random. On second look, one of them might look a little different, the third. The first sequence is just the first 100 decimal places of π: it looks random but it is not. The second sequence is one I generated on my computer using the computer algebra program Maple; and though these are not true random numbers but "pseudo random," they should show no patterns even on a second look. The third sequence I simply wrote down by hand, so it may well show some patterns—even though I was *very* careful.

Indeed, my careful effort to randomize may well have been a mistake. In fact, one can determine that numbers written down by hand always have too many regularities and noticeable problems—sufficiently many, and sufficiently idiosyncratic that one can even tell *who* wrote the sequence down. Marc-André Schulz and Nils Asmussen, two students from the city of Kiel in Germany, investigated this. Their analysis for the competition Jugend forscht ("Young People Do Research," the annual German National Science Fair) showed that one can identify people quite reliably from the random numbers they write down. One person might tend to repetitions, another to numbers that increase, or to alternate between large and small numbers, and so forth. The numbers in my sequence above will therefore have a different character than the set you wrote down.

Is this a game? Is it interesting? Is it important? Well, yes, it turns out to be quite important—sometimes billions depend on that sort of investigation, as well as some serious politics. One can, for example, see if the numbers in the balance sheets of a corporation are as random and as evenly distributed as they should be or if there are some noticeable deviations. When there are deviations, one can always find them. If one reviews the exact vote counts in the disputed Iranian parliamentary election of June 2009, one sees such deviations. For example, one may assume that the last digits of the total number of votes for each district should be pretty random; that is, independent of the size of the district or the number of voters and so forth, each of the ten digits should appear with equal probability. However, they don't. The number 7 appears as the final digit in some 17% of all the results, the number 5, on the other hand, in only 4%, or hardly at all. Strange. If one analyzes other elections in this way, no such aberrations appear. Could it be that someone sat down and invented some "random" results that favored one of the candidates—a candidate who needed an overwhelming majority to claim legitimacy? The method of the young researchers from Kiel might even allow us to determine *who* in the Interior Ministry invented the vote counts.

Numbers can disclose even more. We can again look at the Iranian election results and consider the leading digits. The published results for candidate A in the 366 election districts range over several orders of magnitude, from ten thousand to over a million. For such a range one would *not* expect all digits, 1, 2, 3, ..., to appear with equal probability. When a set of numbers ranges over several orders of magnitude and is not uniformly distributed over a fixed interval but involves percentage increases or multiplications, one expects a distribution of first digits that follows Benford's law. The law was published by the Canadian astronomer and mathematician Simon Newcomb in 1881, but it is now commonly known by the name of the physicist Frank Benford, who rediscovered it in 1938. According to Benford's law, 1 appears as a leading digit in roughly 30% of the cases, 2 appears in another 17.6%, 9 in the smallest fraction, and 0 of course not at all. Why should this be so? For example, if we keep multiplying by the same number over and over, we find that we stay in the range from 1000 to 2000 for much longer than from 2000 to 3000, and so forth. Let's try it. We start with 1000 and add 20% (or multiply by 1.2). Rounding off to integers, we get the following sequence:

1000, 1200, 1440, 1728, 2073, 2487, 2984, 3580, 4296, 5155, 6186, 7423, 8907, 10,688, 12,825, 15,390, 18,648, …

in which 1 appears as the leading digit four times, 2 appears three times, the others only once, and 9 not at all before we get to the four-digit numbers, with leading digit 1 again. If we had started with another number instead of 1000 and used another increase instead of 20%, we would still get a similar picture, as you can verify with a pocket calculator, a spread-sheet program, or a piece of paper.

The powers of 2

1, 2, 4, 8, 16, 32, 64, 128, 256, 512, 1024, 2048, 8912, 16,384, 32,768, 65,536, 131,072, 262,114, 524,288, 1,048,576, 2,097,152, …

show the same behavior: 1 is the most frequent leading digit (in about 30% of the numbers) and 9 is the least frequent (about 5%), but all numbers appear as leading digits. It is just the same for the Fibonacci numbers, where each is the sum of the two preceding numbers:

1, 1, 2, 3, 5, 8, 13, 21, 34, 55, 89, 144, 233, 377, 610, 987, 1597, 2584, 4181, 6765, 10,946, 17,711, 28,657, 46,368, 75,025, 121,393, 196,418, 317,811, 514,229, 832,040, 1,346,269, …

Benford's Law is a very elegant tool in the search for manipulated numbers. Only a few days after the Iranian election, the French mathematician Boudewijn Roukema published a mathematical analysis of the election results on the Internet and submitted it to a professional journal. According to his analysis, there is a very great probability that the votes for the incumbent president were drastically "corrected" up.

This is, of course, not the only example of manipulated results. The accounts of Enron, the energy corporation that went bankrupt in 2001, did not stand up to an analysis using Benford's Law. The books contained too many numbers with leading digits 7, 8, and 9, too few with 1. A similar analysis of the Clintons' tax returns for 1977–1992, on the other hand, showed no evidence of manipulation. In Germany, the numbers that people fill into their tax returns are examined with a Benford Law probe by the various finance ministries (federal and state). Time and again that leads to interesting results.

It is indeed difficult to invent random numbers.

Everything Far Above Average

Garrison Keillor, the radio journalist, reports frequently and thoroughly from Lake Wobegon, a lovely, but thoroughly average, town in Mist County, Minnesota, on the edge of Lake Wobegon. The town is small, with only a few hundred inhabitants (942, to be precise), and there's very little out of the ordinary to report. The general store, Ralph's Pretty Good Grocery, is hardly worth talking about, nor is the statue of the Unknown Norwegian in the square in front of the church. Not much happens in Lake Wobegon, and that is the way the people who live there like it. Sundays they go to church (Our Lady of Perpetual Responsibility); their world is well ordered. In Lake Wobegon, all the women are strong, the men are good looking, and the children are above average—all of them.

All of them above average? That's not possible, at least in the real world, without resorting to statistical tricks. And, in fact, if one keeps one's eyes open, one is confronted with impossible, unlikely, or implausible tricks performed with statistical averages on all sides.

In any case, we can say things in Lake Wobegon are definitely better than here in Germany. If one asks random pedestrians (or journalists, or politicians) how they did in school, one usually gets the questionable answer, "Oh, I was always bad in math" or maybe the less definitive version that they were below average in school. (German chancellor Gerhard Schröder provided, unasked, the same information at a ceremony honoring the prize winners of Jugend forscht. The current chancellor can't claim this, since she did very well at a Mathematics Olympiad.)

Can we believe that nearly everyone on the street was below average? Or is above average in Lake Wobegon? Well, yes: if everyone in a class, with one exception, has an F, then the class average is a tad better than F, and everyone in that class—with one exception—is indeed below average.

We may draw several different conclusions from this example. One certainly is that in very skewed distributions, the average may have only a limited usefulness. For another example, what does an average income tell us? Note that if a financial speculator in Germany, say, gets a windfall of €80 billion, the average income of the 80 million German citizens rises by €1000, even though none of them sees any benefit (until, perhaps, the stock speculator pays his taxes, which may be an unlikely event).

In fact, for many purposes the so-called median is a much better and more useful quantity than the average. One computes a median of a set of numbers by putting them in order, smallest to largest; the median is the number in the middle. (If we have an odd number in the set and if all the numbers are different, the result is unique and unambiguous.) Considering the distribution of incomes again, half the population has an income below the median and half has an income above the median. To put it another way, the median divides the population into two halves, one with incomes below the median, the other incomes above. If one of the peak earners makes a few millions more or less, that does not change the median. And in the class of students we considered earlier, the median is an F no matter what the one better student got.

Medians are also used to determine what fraction of a population lives in poverty. The usual definition is that anyone with a disposable income less than half the median income is poor. This may be a practical definition, but it has some snags. It says nothing, for example, in absolute numbers about what income level is needed for anyone to survive, it only considers relative income levels. So if suddenly everyone receives twice as much income as before (admittedly quite unlikely), the number of people counted as being in poverty does not change. Ditto when the incomes don't change but all prices are doubled.

We should notice another statistical anomaly with this definition of poverty: the number of "poor" people can never be more than half the population—it is mathematically impossible. If Germany, for example, had a few billionaires while everyone else received nothing more than long-term unemployment benefits, the median income would be determined by the dole, and no one would be "poor." If everyone got a raise of a million, no one would still be poor. Is this reasonable? I don't think so—even if "riches for everyone" is always a popular demand (on the left) or promise (on the right). One particularly impressive variant was the main slogan for the conservative CSU party in Germany during the spring 2008 federal campaign: "More net income for all." Income redistribution may be able to change the median income or the overall spread of incomes, but it can do nothing to change the average or the total sum of all incomes. Most such campaign slogans are to be regarded with mathematical skepticism.

Integers

Recently the German version of *Who Wants to Be a Millionaire?* featured a rather clueless candidate who finally asks the audience for the answer. The host's screen gives the correct answer as provided by the audience: C, "with 96% certainty." That surprises the candidate, who says, "Oh, I should have maybe known that." The host tries to reassure him: "Maybe only five people voted."

What's wrong with that? Well, if only five people voted, then four people might have voted for C, so that would be 80%, or if all five voted for C, it would be 100%; the result cannot be between these (unless we have one very schizophrenic audience member). If the result rounds off to 96%, there have to be at least twenty-three participants in the vote (if one of the twenty-three votes is for something other than C, then 95.65% are correct, which rounds off to 96%). To get a result of exactly 96% (which, of course, no one claimed), some multiple of twenty-five people have to be voting.

"Recognize numbers" is our motto here. In order for a poll to produce a result of 99% (rounded off), it must have had at least sixty-seven respondents. On the other hand, if the poll produces 14.28% for (or against) some option, caution is indicated. That percentage is pretty nearly one seventh, so the whole survey might have included only seven participants, of whom one gave this answer—or was it perhaps as many as fourteen who were asked? Sometimes numbers reveal more than appears at first glance.

One of the keys to recognizing numbers is to consider integers. Surveys, polls, and elections always include whole numbers of participants, halves of people do not take part, and individual people also do not cast votes that say, "I'm two thirds for that." (An exception was a provision in the U.S. Constitution's call for a census that counted each slave as three fifths of a person.) Indeed, there is the biblical injunction "But let your communication be Yea, Yea; Nay, Nay: for whatsoever is more than these cometh of evil." This is from the Sermon on the Mount (Matthew 5:37). Jesus is preaching on the integer results of opinion polls.

We can also make use of our recognizing whole numbers at the supermarket checkout, for example, with the "penny test": I have a dozen objects in the cart, costing between $0.69 (yogurt) and $5.99 (a large container of organic muesli). The cash register shows a total of $22.73. What is wrong? In my supermarket all the prices end in 9, so the individual prices could have been $0.69, $1.29, $2.79, 2.99, or 5.99. Each price falls just one penny short of the nearest dime. For twelve items, then, the total must be 12 cents short of some number of dimes and the sum must therefore end in 8, not 3. The sum of 22.73 could have been for seven items, or seventeen, but not twelve.

If the total *should* have been $9.78, then you should also be taken aback at seeing $22.73 on the register. And don't tell me, "But the items were all scanned, and the computer did the addition, so it ought to be right." That may be true, but one should still be skeptical. Do you really want to trust that all the prices were entered correctly into the supermarket's computer? Or that the register adds correctly? Not I. But that may be because I've been burned once.

Not too long ago, my bank irritated me with a credit card statement that included in bold capital letters "REPLACEMENT STATEMENT OF 09/21/2007—ENTRY DATE OF 09/26/2007 IS INVALID." Of course one is tempted to compare the replacement statement with the "invalid" one. The two statements were practically identical; the same amount brought forward, the same eight posted transactions, but differed in the total sum, and that by a considerable €338.80, and not to my benefit. Such a thing should not occur. I should have noticed myself (but who would suspect that a bank can't add correctly?). Of course I called the bank. Their answer: the sum is indeed correct, but the earlier statement referred to a different closing date, and thus did not account for all the eight listed entries. Well, maybe. But irritating nonetheless. And, anyway, "Take care of the pennies and the dollars will take care of themselves," as the saying goes—hence, the penny test at the supermarket cash register.

Chapter 4

Caution: Equations

"Equations are the language of mathematics." Sounds plausible. Heads nod involuntarily in assent. But when you look at the statement carefully, it's clearly nonsense: no mathematician in the world speaks in only equations. One can read equations, one can write equations, one can even erase equations, but one does not speak them. Equations are not a language but a script.

There is, of course, mathematics without equations, but for mathematicians, as for physicists, engineers, and astronomers, equations are a desirable goal. We are always glad when we can summarize a state of affairs in an equation—a process we call "modeling"—because equations express things precisely and exactly; for equations there is only correct or incorrect, no "possibly" or "something like that."

But also: caution must be exercised when dealing with equations. Not everything that looks like an equation contains something mathematical. Equations are well known to be the hallmarks of mathematics, the way one recognizes it and sees it coming. They are therefore also used to suggest or insinuate that an argument has something of mathematics in it.

Equations for Everything?

Headlines of the sort "Mathematicians Find a Formula for …" are easy to find, especially during the dog days of summer when there's not much with which to fill a newspaper or magazine. When we see something like that, we can allow ourselves to be a little skeptical. One can indeed put lots of things into formulas or equations (even recipes, dance steps, and ways to knot a tie), but simply writing an equation down will hardly solve a serious problem.

When I see such a headline, I immediately ask myself, is this meant seriously, and if so, by whom? After all, the stories we see generally come to us like a game of telephone or Chinese whispers. The message starts with a scientist or research team, is then taken over by the publicity department of the university, transferred to a wire service, reviewed by an editor, assigned to a reporter, and then printed for my benefit as reader. At which of the many transfer points was the message treated seriously? Was serious research treated as a banal effort, or was a mathematician's attempt at an April Fool's joke revived for a filler in August?

See for yourself. Here are three prize examples from my extensive collection of stories about equations.

1—Nine Variables

On 20 June 2008, the *Daily Mail* of London reported:

SCIENTISTS DISCOVER A NEW MATHEMATICAL FORMULA … FOR THE PERFECT CHEESE SANDWICH

Most of us are happy to slap two pieces of bread together with a few slices of cheddar and, if we're lucky, a squirt of salad cream to make a good cheese samie.

But it seems the process may have just become a bit more complicated than that.

Not content with their usual figures and algebra, scientists have discovered a mathematical formula for creating the perfect cheese sandwich.

The volume of mayonnaise or pickle to calculate the ideal cheese thickness to go with the relish is essential, according to scientists.

Sensory analysts at Bristol University have devised an equation into which diners follow factors such as how much mayonnaise or pickle to put in the sandwich and the ideal cheese thickness to go with the relish.

The formula, which includes nine algebraic variables, has been used to create an online calculator, which can be seen at www.cheddarometer.com.

The formula (detailed below) is the result of a study by senior research fellow Geoff Nute and his team at the university's Sensory & Consumer Group in the Division of Farm Animal Science.

Using human assessors and complex measuring devices, the group claims to have "mapped" the flavour profile of hundreds of samples of Cheddar to determine the tastiest thickness for each type of filling.

Mr Nute said: "We used specially trained human taste testers to sample a range of Cheddar cheeses in a carefully controlled environment and combined results from these tests with instrumental data obtained using colorimeters and pressure sensors to obtain precise measurements of variants such as yellowness, crumbliness, creaminess and tanginess.

"The results of our research have been extrapolated to produce a formula which takes into account modifying characteristics of individual cheeses and the ratio of popular fillings and achieves a mathematical balance of flavours in order to gauge the correct thickness of the Cheddar."

Philip Crawford, chairman of the West Country Farmhouse Cheesemakers group, said: "We are very proud of our authentic farmhouse Cheddar which we make by hand on our farms using only milk from our own cows.

"This means each variety has a unique character and we were fascinated to know which combinations of sandwich fillings work best with each cheese.

"Collaborating with Mr Nute and his team we have managed to create the Cheddarometer and reveal the blueprint to everyone's perfect cheese sandwich."

Last year the collective turned a traditional round of Cheddar into an unlikely internet hit, when 1.8 million visitors looked at a webcam showing the cheese maturing in real time.

The formula:

$$W = \left(1 + \frac{bd}{6.5} - s + \frac{m - 2c}{2} + \frac{v + p}{7t}\right)\left(100 + \frac{l}{100}\right)$$

W = The thickness of Cheddar in millimetres
b = The thickness of the bread
d = The dough flavour modifier
s = The thickness of margarine or butter
m = The thickness of mayonnaise
c = The creaminess modifier
v = The thickness of tomato
p = The depth of pickle
t = The tanginess modifier
l = The thickness of the lettuce layer

The version of the story that I saw appeared a day later in the *Hamburger Abendblatt*; it was a little less explicit:

London—Britische Forscher haben eine mathematische Formel für die perfekte Herstellung eines Käsebrotes gefunden. Dabei komme es auf die richtige Dosierung von Mayonnaise, Salat und Cheddar-Käse an, sagte der Leiter der Studie, Geoff Nute (Universität Bristol).

Für Kenner: Es handelt sich um eine Gleichung mit neun Variablen (im Internet: www.cheddarometer.com). Die Matheformel:

$$W = \ddot{A}1 + \left(\frac{bd}{6.5} - s + \frac{m - 2c}{2} + \frac{v + p}{7t} \right) \ddot{U} \left(100 + \frac{l}{100} \right)$$

W ist die Dicke der Käsescheibe in Millimetern, *b* die Dicke des Brotes, *d* seine Besonderheit, etwa Weißbrot oder Vollkornbrot. Die anderen Variablen stehen für die Salatmenge (l), Gewürzgurken (p) und Tomaten (v). Die Formel wurde mit menschlichen Versuchskandidaten, komplizierten Messgeräten und mehreren Hundert Cheddarsorten ermittelt.

Which makes the equation for *W* even less transparent. On closer inspection, we see that in Hamburg the conversion of the press release to the newspaper's typesetting program changed the square brackets of the original to Ä and Ü. But even the *Daily Mail*'s formula cannot be correct, since there are more closing parentheses than opening parentheses. And if we actually count the variables, we see that there are ten, not nine: *W*, *b*, *d*, *s*, *m*, *c*, *v*, *p*, *t*, and *l*. Apparently what was meant was that it takes nine variables to determine the tenth, *W*. But it is still not clear what units of measurements were used for the different quantities (are they all millimeters?), or how the "modifiers" *d*, *c*, and *t* are even to be measured. Mr. Nute may know, but the reporter did not tell us.

2—Sexy Shoes

The next example comes from *Spiegel Online*, the online version of a German newsmagazine. On March 22, 2004, one Markus Becker reported on "A Formula for Sexy Shoes":

> How tall can the highest heels be? A British researcher called upon mathematics and has found a formula with which women can compute the maximal height for the heels of their shoes.

The story continues and provides the formula:

$$h = Q \times (12 + 3s/8)$$

where *h* is the maximal height of the heel (in centimeters), *s* is the shoe size (the British shoe size), and *Q* is a "sociological" factor that lies between 0 and 1. Indeed. We then learn that *Q* can itself be computed according to a formula:

p ? (*y* + 9) ? *L* divided by (*t* + 1) ? (*A* + 1) ? (*y* + 10) ? (*L* + 20£)

where *p* represents a "sex-value" of the shoe on a scale from 0 to 1, and so forth.

Before we even get to the interesting question of who determines the sex-value of the shoe (and how do they determine it?), we should wonder, what do all those question marks mean? Could they have been multiplication signs that somehow got changed in the transfer to HTML for the web page? Did nobody (perhaps including Mr. Becker) realize that a question mark does not represent a mathematical operation? Or maybe this is not important for the story. In any case, this story is still available online. In 2009 the question marks were still there, but they have now (in 2012) been replaced by dots, a common mathematical symbol for multiplication. The story is filed under the category "Science," subcategory "Man and Technology." It is illustrated with photos of Kylie Minogue, Madonna, and Marilyn Monroe, all wearing high heels, of course.

3—Horror Math

Another story. On August 16, 2004, *Spiegel Online* reported that British mathematicians "have now developed a formula for the perfect horror movie." The BBC had reported the same result a few days earlier:

> The secret of making a scary movie has been calculated by university experts.
>
> Scientists have worked out an equation to prove why thrillers like *Psycho* and *The Blair Witch Project* are so successful at terrifying audiences.
>
> The formula combines elements of suspense, realism and gore, plus shock value, to measure how scary a film is.
>
> Researchers spent two weeks watching horror films like *The Exorcist*, *The Texas Chainsaw Massacre*, and *Silence of the Lambs* in pursuit of the formula.
>
> The model focuses on three major areas: suspense, realism and gore.
>
> **Shock impact**
>
> Factors considered include the use of escalating music, the balance between true life and fantasy, and how much blood and guts are involved.
>
> As suspense plays such a pivotal role in the success of a scary film, its elements—escalating music, the unknown, chase scenes and a sense of being trapped— are brought together and then squared. Shock value is then added.

SCARY MOVIE FORMULA

$(es + u + cs + t)$ squared $+ s + (tl + f)/2 + (a + dr + fs)/n + \sin x - 1$.

where:

es = escalating music
u = the unknown
cs = chase scenes
t = sense of being trapped
s = shock
tl = true life
f = fantasy
a = character is alone
dr = in the dark
fs = film setting
n = number of people
sin = blood and guts
1 = stereotypes

What the x is whose sine represents blood and guts is not said. Nor is it clear how the 1 for stereotypes is to be treated when there are no stereotypes in the movie.

Not always is the nonsense presented as "mathematical formulas," as obvious as it is in these three examples. Obviously, if the parentheses don't match up, or if funny symbols appear, the equations cannot be proper mathematical relationships. But this is also true if the variables that appear in the equations have no clear meaning, or if it is not specified how they are to be measured or what units are to be used. Equations promise clarity and precision, but they can also dissemble: specious "sociological factors," "correction terms," and the like should alert us to possible problems.

A few other indicators are also noteworthy: these stories about mathematical formulas in the popular press often involve sex or violence. Sometimes the sexism is incorporated in the formulas directly, as in the story about Chinese researchers who found a single formula for the appeal that female (!) bodies have for male (!) observers. This appeared in *Spiegel Online* on the 14th of January 2004.

That it is often British researchers who are the sources of the serious German stories about mathematical nonsense might suggest that German newspaper editors don't understand British humor. One should perhaps develop a formula for this.

But before I work on that, I need to resist the temptation to develop a formula for determining nonsense in equations, a formula that I would, of course, send to the publicity office of my university in time for the press release to go out on the first of April.

The Body Mass Index

The proverbial man on the street (and even more so, the woman on the street) doesn't seem to feel much practical need for the Pythagorean Theorem but is convinced that the stars determine much of our lives, and that the BMI, the Body Mass Index, controls our medical fate.

In case you are not familiar with it, the BMI is given by the formula

$$BMI = Weight/(Height)^2$$

where the height is measured in meters and the weight in kilograms. (If the weight is measured in pounds and the height in feet, there is a conversion factor of 4.88 that needs to be included in the formula.) This is not really an equation but rather a definition that provides a number, the BMI, that

may or may not be useful. The definition was first put forth
by a Belgian mathematician and statistician named Adolphe
Quetelet (1796–1874).

The BMI, which is also known as the Quetelet Index, can
be computed for anyone. It is then supposed to have a simple
interpretation: people with a BMI below 18.5 are considered
underweight, according to the World Health Organization; a
BMI between 18.5 and 25 is considered normal; people with a
BMI between 25 and 30 are considered overweight, and those
with a BMI over 30, obese. The ranges have been adjusted
from time to time—in 1998, for example, the U.S. National
Institutes of Health dropped the upper bound for normal
weights down from 27.8 to 25, matching the WHO scale,
and instantly rendering millions of U.S. adults "overweight."
The fact that most of the limits fall conveniently on multiples
of five suggests that the limits were set not for statistical or
medical reasons, but because they are easy to remember and
to publicize. In addition, the medical usefulness of the BMI is
clearly limited since it cannot distinguish muscle from fat.

But since we have the formula here in front of us, let us
look at it more carefully. People are mostly water, so we can
suppose that their weight is pretty nearly proportional to their
volume. The BMI is thus pretty nearly a geometric quantity:
volume divided by height squared. But what does that mean?

In considering that, I'm reminded of the famous case of an
unrealistic geometry problem: "Let K be a spherical cow...." Of
course, the volume of the cow is then easily computed as $(4/3)\pi$
times the radius of the cow cubed. Quite exact, but not realistic.

So how does the BMI say anything about obesity? Well, we
can fit a model to the formula and see what happens. Let us
imagine an adult as a block—a bit like a Lego person—the
volume of our model person is then simply height times width
times thickness; or, if we assume the width is ¼ of the height,
then it is ¼ h^2d, where h is the height and d is the thickness,
front to back, of the block-person. When we compute the BMI

of our block-person, the square of the height in the denominator cancels that in the numerator, and the BMI is simply proportional to the thickness of the person; again, since a person has
pretty nearly the density of water, a bit of multiplication and
keeping track of units leads to a BMI that is 2.5 times the front-
to-back thickness measured in centimeters of our block-person.

You're not convinced? You don't think the block makes
a realistic model? Well, I don't either. It's simply a geometrical model that allows for an easy interpretation of the BMI.
Quetelet certainly had no such geometrical model in mind
when he developed the BMI. Instead, he diligently collected
statistical data, including among others 5,738 Scottish soldiers.
Next to the actual physical measurements of his subjects,
Quetelet also noted which of them looked like they were overweight or underweight. Once he collected all the data, he saw
that the BMI index was quite nicely correlated with his nota-
tions. So the index is based on statistics, not geometry.

In this context we should look again at the story from
Spiegel Online about the researchers who found a formula for
sex appeal:

> What makes women attractive to men? Chinese
> researchers are said to have found the solution. In
> their opinion the attractiveness of a female body can
> be explained with just one simple formula....
> The formula is: divide the volume of the body by
> the square of the height of the body measured from
> the feet to the chin. According to the research, men
> can see the result intuitively and can tell on the
> first glance.

Well, now, we have already supposed that body weight and
body volume are pretty nearly proportional, and we can also
be pretty sure that the height of a human measured to the
chin is very well correlated with the height measured to the

top of the head. So the magic Chinese formula is essentially nothing more than the BMI. That should not have warranted a press release.

Postscript

Of course I wanted to try out one of the BMI computers on the Internet. My actual weight should not be of concern to the Internet, so I just typed in:

Height 185 cm, weight 180 kg

Answer

Your BMI value is 52.6
Your weight is clearly above the normal range. In the long run, this can lead to disease. You should have a thorough medical exam and discuss with your doctor how best to reduce your weight. In general the most successful methods for this are to reduce your food intake and increase your level of activity.

That seems appropriate. A little further experimentation showed that the online BMI computer accepts weights between 39 and 250 kg (85 and 550 lbs.). In case your weight is outside those limits, it's probably a good idea to see a doctor in any case, with or without a computed BMI.

The Huntington Affair

The American number theorist Neal Koblitz is known to specialists because of several quite advanced treatises on number theory as well as an important cryptographic technique, based on hyperelliptic curves, that he published in 1989. But already in 1981 he had planted a time bomb that took several years to detonate: he published an article, "Mathematics as Propaganda" (in *Mathematics Tomorrow*, L. A. Steen, ed., Springer-Verlag, New York, 1981), that included, among others, a polemic against publications of Samuel P. Huntington.

Huntington (1927–2008) was an influential American political scientist, a professor of government at Harvard University, and an advisor to U.S. government agencies (including the CIA) under several presidents, as well as a consultant to the governments of Brazil and South Africa. In 1993 he published an influential article, "The Clash of Civilizations?" in which he predicted that with the end of the cold war conflicts would be based on cultural differences, in particular between Islamic and Western cultures. The article (which Huntington later expanded into a book) led to lively debates and may well have had a strong influence on George W. Bush's fatal foreign policy.

In an earlier book, *Political Order in Changing Societies*, published in 1968, Huntington sets up the following formulas (on page 55, in case you want to look them up):

(1) $\dfrac{\text{Social mobilization}}{\text{Economic development}} = \text{Social frustration}$

(2) $\dfrac{\text{Social frustration}}{\text{Mobility opportunities}} = \text{Political participation}$

(3) $\dfrac{\text{Political participation}}{\text{Political institutionalization}} = \text{Political instability}$

What is that supposed to mean? One could read these formulas as qualitative relationships, indicating, for example, that social frustrations rise as social mobilization rises while economic development stagnates. That, at least, is plausible. (This is not a topic we can discuss here, as we would need to include sociologists and economists in the discussion.)

But what was written on page 55 were not just general relationships but actual equations. Indeed, Huntington referred to a study published in 1966 in which these equations, with sets of numerical values for the variables for the conditions in Belgium, France, and South Africa (all mentioned in the same breath), were used to show that these three were all "satisfied societies," but that they showed a high degree of "political instability." France unstable? South Africa satisfied? In 1966? And proven mathematically? Let's look at the source cited by Huntington. This is the article by Ivo Feierabend and Rosalind Feierabend, "Aggressive Behavior within Polities, 1948–1962: A Cross-National Study," published in volume 10 of the *Journal of Conflict Resolution* in 1966, and which, by the way, won the "Socio-Psychological Prize" of the American Association for the Advancement of Science in the year it was published. In that paper, Ivo and Rosalind Feierabend establish the following formula:

$$\frac{\text{Social want satisfaction}}{\text{Social want formation}} = \text{Systemic frustration}$$

This seems plausible only if you let yourself be intimidated by the equation. Wanting much and getting little leads to frustration, as any spoiled five-year-old can tell you on Christmas Day. But that is not what the formula says: if the denominator is large and the numerator is small, the result of the division is also small. The indicated fraction thus provides a "frustration index" that becomes small when the frustration is large.

But this "frustration index" is not just a qualitative indicator for use at Christmas. No, it is supposed to provide rigorous numerical results. Feierabend and Feierabend and their colleagues collected and presented data that provided measures of wish formation and wish fulfillment using the same scale for a wide variety of countries:

The frustration index was a ratio. A country's combined score on the six satisfaction indices (GNP, caloric intake, telephones, physicians, newspapers, and radios) was divided by either the country's coded literacy or coded urbanization score, whichever was higher.

A few pages farther on in the paper by Feierabend and Feierabend, we see the same formula—but now with a rule for evaluating it:

> In the first hypothesis, the discrepancy between social wants and social satisfactions is postulated to be the index of systemic frustration. The relationship is represented as follows:

$$\frac{\text{Want satisfaction low}}{\text{Want formation high}} = \text{High frustration}$$

If one reads on, it only gets worse:

$$\frac{\text{Want satisfaction low}}{\text{Want formation low}} = \text{Low frustration}$$

$$\frac{\text{Want satisfaction high}}{\text{Want formation high}} = \text{Low frustration}$$

Do you notice something fishy in this arithmetic? If the denominator of a ratio is large and the numerator is small, one can surely conclude that the ratio will be small (not large!), but if both denominator and numerator are large (or both small), one can conclude nothing about the ratio without further information. This is, what, fifth-grade arithmetic?

I claim (and in this I join the critique by Neal Koblitz) that the use of such "formulas" in the works of Huntington and of Feierabend and Feierabend is in fact dangerous: here mathematical methods are used in a specious and nonsensical way to produce results that are then used by others for, in the end, political results (including U.S. foreign policy).

Why is mathematics misused in this way? Why formulas and equations? Because this gives the discussion an aura of exactness, of unassailability. Koblitz expressed it this way: "Huntington's use of equations produced effects—mystification, intimidation, an impression of precision and profundity."

This, then, was Koblitz's time bomb of 1981. A few years later, it was detonated. In 1986 Huntington was nominated for membership in the very prestigious National Academy of Sciences of the United States. Although the nominations are made by committees from each of the disciplines represented in the National Academy, it is the entire membership that has to vote in the election. In general, this is completely routine and goes without further ado. In Huntington's case, however, there was a protest. Serge Lang, a highly renowned professor of mathematics at Yale, author of dozens of textbooks and technical monographs, and a highly disputatious person, had just been admitted to the academy the year before. Lang accused Huntington of grievous misuse of mathematical methods, referring to Koblitz's "time bomb." He proceeded to write hundreds of letters, both public and private, to mobilize supporters and organize a veritable campaign against the election, with the result that Huntington was not elected to the academy. Huntington's supporters did not want to let this disgrace stand and nominated Huntington again in 1987. Another bitter contest led to another failed election. Conclusively.

Huntington, with his equations that are not equations, attempted to suggest precision and gravitas for his arguments—but these specious (pseudo-)mathematical arguments instead made him blameworthy. He was publicly exposed and suffered for it. One can feel some satisfaction in that, no?

Postscript

The Russian poet and novelist Leo Tolstoy was proud of his rule of fractions: men are fractions, he said, in which the man's soul, his best qualities, and true merit are the numerator and his opinion of himself is the denominator. In interviews, he said that he had evaluated his colleagues and competitors using this measure. The decadent schools of literature in Russia and the West—he named Henrik Ibsen, Oscar Wilde, and others—had such great opinions of themselves that the ratio was reduced to nil. Though Tolstoy probably never wrote his rule as a formula, we can do so here:

$$\frac{\text{Achievement}}{\text{Self-esteem}} = \text{Worth of a man}$$

Pythagoras Lives

The theorem of Pythagoras is probably the most famous equation of mathematics, and also has all the necessary characteristics (about which we need to speak shortly). However, the form in which it is usually presented:

$$a^2 + b^2 = c^2$$

is totally abstract and also totally meaningless.

Well, do *you* know what it is supposed to mean? Not I. Indeed, the formula has no meaning without the information about what the letters a, b, and c are supposed to refer to. Of course, in this case Pythagoras tells us that they refer to the lengths of the sides of triangles with one right angle, with c being the longest side, the one opposite the right angle.

And who is supposed to care about that? The fans of right triangles, perhaps? There can't be very many of those, one would think. So then why would Pythagoras be making up theorems for such a fringe group? Wikipedia tells us that the theorem is one of the fundamental theorems of Euclidian geometry, which may make it a little less of a fringe interest. If it still seems excessively academic, it really is not. The

Pythagorean Theorem is nothing less than the key to survey-
ing and measuring the world. It provides the link between
geometry and measurement. Solved for c in terms of the two
other sides:

$$c = \sqrt{a^2 + b^2}$$

it tells us how to compute distances.

That this is interesting not only in the abstract but also in
practical cases was brought home to me a few summers ago
as we tried to put an old wardrobe into the attic above my
mother-in-law's new apartment. To carry it up the stairs and
through the door we planned to put it on its side. But then the
question arose, could we stand it back up again once it was in
the attic? Was the ceiling perhaps too low? The wardrobe was
120 cm wide and 160 cm high; its diagonal was thus (according
to Pythagoras) 200 cm. The ceiling was just a tad over 200 cm
high. Fortunately, then, no problem—just a little surprise to
meet Pythagoras in my mother-in-law's attic. And an even big-
ger surprise that the arithmetic could proceed so smoothly and
exactly without benefit of a calculator. That was due to the
measurements of the wardrobe: the ratio of 120 cm to 160 cm
is just 3 to 4, and the famous sum $3^2 + 4^2 = 5^2$ was already to be
found in Babylonian clay tablets, long before Pythagoras.

That we could use the Pythagorean Theorem in my
mother-in-law's attic is not only due to my mathematical perspective but also due to the universal validity and occasional
applicability of the theorems of mathematics.

The Pythagorean Theorem is indeed not just any old
formula, but is the very exemplar of an equation of mathematics. There are clear rules and characteristics of such an
equation. The most important of these are clear premises for
the statement and a clear range for its validity. Once these are
given, the formula is exact and exceptionless. The formula
$c = \sqrt{a^2 + b^2}$, after all, does not apply to any three numbers

a, b, and *c,* but only to the lengths of the sides of right triangles, with *c* the length of the longest side.

Because the formula holds exactly, it can be solved for each of its relevant components, again following clear and universal rules. One can thus compute not only the longest side given the lengths of the two others, but one can compute the length of either short side from the remaining sides. One can insert such formulas into further computations, combine them with other true formulas, and continue the calculations. What comes out of that may not always be interesting, but it will always be true and will always hold exactly. But I can guarantee that if you use poor estimates for two of the numbers and compute the third, it won't produce a useful result, in spite of the rules of mathematics.

One could have argued that the Huntington equations and the Feierabends' equations are only approximately valid. But we cannot let that evasion pass. If someone writes an equals sign and calls his equations "equations," he must allow them to be judged as equations. The equals sign is a signal, a litmus test: it says, there is mathematics here. Someone who uses it enters the domain of mathematics. The equals sign is, after all, not a part of everyday discourse; it is an invention of mathematics, and a comparatively recent invention, at that. It was introduced by the Welsh doctor and mathematician Robert Recorde (1510–1558) in his book *The Whetstone of Witte* in 1557, in which he intended to sharpen the reader's wit in the use of mathematics and effectively introduced algebra to England. Recorde designates two parallel lines to represent equality, "bicause noe 2. thynges, can be moare equalle." Someone who writes "=" should therefore mean it. If one means "behaves like" or "is approximately" or "corresponds roughly to," one should say so and not use the sign that says "no two things can be more equal."

Subreption of significance through mathematics—that is, the use of mathematical expressions to give specious import to a statement—is actually quite common. In addition to equals signs, right triangles are frequently misused in this way. Astrology makes much use of these: geometric diagrams with planetary positions, dates, and much else are festooned with squares, rectangles, trapezoids, mirror axes, acute and right triangles, congruent figures—either precisely aligned or perhaps missing by a degree or two or three. And, of course, one can always find something that one can interpret as having significance using the magic cloak of mathematics and geometry. The triangles and such of course have mathematics in them—but my fortune?

Postscript

If it is esoteric, then it is immediately correct. I recommend *Dirk Gently's Holistic Detective Agency* by Douglas Adams, not only for the Electric Monks, robots made to believe things for you, thus saving you from having to believe them yourself, but also for the sofa in the stairwell, which already *theoretically* cannot fit there and which one therefore cannot remove from there, for decades.

Equations as Art

Craig Damrauer is an artist living in New York who has been designing "New Math," which he defines as a "somewhat quixotic attempt to quantify the world." When one sees a statement like that, several alarms should go off: first, we have again an attempt to pack everything into mathematical formulas, and, second, the naive search in physics to stuff everything into a single equation that will explain life, the universe, and everything else.

But Craig Damrauer is not a physicist, nor is he a mathematician. He is an artist. His New Math equations are, I believe, meant to be taken seriously. Here are three:

$$\text{DIVORCE} = \frac{\text{MARRIAGE}}{2}$$

$$\text{CHOCOLATE} = \frac{\text{MAKES HAPPY}}{\text{MAKES FAT}}$$

$$\text{ANGER} = \text{TICKED OFF}^3$$

Each Monday, Damrauer publishes a new equation on the website www.morenewmath.com. Are these really equations? When we look at the chocolate formula (more chocolate should mean more fat, no?), we should say no, that's not math. But perhaps an artistic license overcomes the strictures of the equals sign.

In an interview published in *Jetzt*, a youth supplement to the newspaper *Süddeutsche Zeitung*, Damrauer said:

> I often think in metaphors and similes. And at some point I realized that metaphors work in principle like equations. I find that this is a calming way to describe the world.

In the same interview, he was asked if everything between heaven and Earth could be packed into a set of equations:

> If you ask a physicist or mathematician that question, she would presumably say, yes, everything can be explained with mathematics. But I don't think that that is true for New Math.

One can't add much to that. Equations are not just a way of expressing mathematics, they also expand the language of the arts. An artist can play with equations and use them to express meanings that one cannot grasp in other ways. And that is also one of the great hopes of mathematics.

Chapter 5

The Small Puzzles

"Just what is it that you do?" The question frightens me. I am a mathematician. I am proud of my work, my skill at solving problems, my intuition. But just what is it that I do?

In mathematics there are the Big Problems, the famous and important problems, the problems for whose solution million-dollar prizes have been offered, the problems listed by Hilbert in 1900, the Millennium Problems of 2000. I won't be solving any of these.

There are also smaller problems in mathematics. Problems that may not be vital in the progress of mankind and may not even be vital to the overall structure of mathematics as a whole, but that are pretty, challenging, and stimulating. Problems that one can understand without a specialized and thorough study of mathematics. But I have no idea how one would go about solving these problems.

And then there are the problems of the sort that I am working on. Problems of geometry, concerning polyhedra in particular, that pose rather specialized questions about particular details that lie on the path to a broader and deeper understanding of geometry. The individual problems and their solutions may in themselves not be so important, but they contribute to development of methods, and they provide small building blocks for the larger theoretical structures. These structures in turn provide new perspectives for tackling yet other problems. In this way the circle of knowledge grows ever wider.

Sudokus

At breakfast I read a newspaper, *Tagesspiegel*. Sometimes I find the front of the paper irritating (especially on Mondays, when there is an arrogant know-it-all columnist pontificating on the front page) and seek refuge on the page that has two Sudoku puzzles on it. I turn to the one on the right (labeled "difficult"). In that, I am probably in very good company. Even if only one German in a hundred spends a quarter of an hour working on a Sudoku puzzle, this adds up to some 200,000 person-hours a day. I claim (and only partly as self-justification) that this is not wasted time but rather brain-cell training of the best kind. And the best part is that these 200,000 person-hours are spent "doing math"—completely voluntarily and (one hopes) with pleasure and satisfaction.

Is that really math? Of course it is. Even if the newspapers don't understand that or try to hide it from their readers if they do understand it. The London newspaper *Independent*, for example, introduced Sudokus in 2005 with the introduction:

> Fill in the grid so that every row, every column, and every 3x3 box contains the digits 1–9. There's no maths involved; you solve the puzzle with reasoning and logic. *Stay Sharp*.

Question: why should "thought and logic" be something different from mathematics? When Sudokus arrived a little later in German newspapers, the *Tagesspiegel* soon also presented them—"The cult puzzle finally in Berlin"—and introduced them with the standard clichés:

> It is noteworthy that this is indeed about numbers, but not at all about mathematics. Those who think back with dread on logarithms and derivatives can participate here, as can those who hate crossword puzzles.

What is noteworthy is that the *Tagesspiegel* here missed the mark not once but twice. Sudokus are of course mathematical but have no intrinsic connection with numbers. One could just as easily make Sudoku puzzles with the nine different letters A E F G I L P S T from "TAGESSPIEGEL FAILS" or with any other set of nine symbols (colors in a stained-glass window, for example). Of course, the nine numerals are a practical set to use since they are so familiar that it is easy to see at a glance if one is missing. Furthermore, the "cult puzzle" had already been in Berlin some time earlier. While he was at the court of Frederick the Great, Leonhard Euler had some ideas in this direction: in 1792 he investigated square arrays in which each row and each column had the same elements without duplications. One knows them today as Latin Squares or Euler Squares.

The common 9-by-9 Sudokus can be difficult for us to solve at the breakfast table. For a computer, however, such a puzzle is a trivial matter: they can easily be programmed to solve them in fractions of a second. But one can also consider larger Sudokus, maybe 16 by 16 or 25 by 25, or, more generally n^2 by n^2. And for these it has actually been proven that the solution is difficult—proven, that is, if the conjecture that "P is not NP" is correct, a conjecture that is widely assumed to be true but remains one of the really great, famous, and important unsolved problems of mathematics, about which I will say more later.

Puzzle aficionados might be interested in knowing how many different Sudoku puzzles there are—for some of them,

out of concern that they might run out of new material for
their pastime at some point. Answering that question turns out
not to be easy. In general, it is agreed that a proper Sudoku
has a unique solution, not several solutions. And if there is
to be a puzzle, then at least a few of the numbers must be
left blank, no? In the introduction to their Sudokus, *Die Zeit*
answered our question in this way: "The number of possible
puzzles is given by the *Herald Tribune* as 10 raised to the 50th
power—an enormous number with 49 zeroes."

Very clever of *Die Zeit* to quote the *Herald Tribune*.
And the number cited is impressive, as 10^{50} is an enormous
number with not just the 49 zeroes but also another one
not counted by the newspaper. But just because a number
is impressively large does not mean that it is impressively
true. Do we know where it comes from? Is it computed or
just guessed? I don't know. It is at least easy to compute the
number of distinct *solutions* of Sudoku puzzles—that is, the
number of different ways to fill a 9-by-9 grid with the digits
1 through 9 in such a way that each row, each column, and
each 3-by-3 square contains each digit just once. There are
exactly 6,670,903,752,021,936,960, or about 7 quintillion. But
that is the number of solutions, not the number of puzzles.
How many ways are there of leaving numbers out of each
solution to turn it into a puzzle? Is the number of actual
puzzles then the 10^{50} cited in the paper, or perhaps a smaller
number, closer in magnitude to 10^{20}? Was the paper citing a
computation or just a guess? I don't know.

But it is actually another unsolved problem that runs
through my head as I'm working on my Sudoku fix in the
morning: how many entries of the grid must be filled in by the
puzzle maker in order to make a puzzle with a unique solution. There are many puzzles with just seventeen entries. The
Australian mathematician Gordon Royle has collected 49,151
distinct ones of these (up to December 2012). Figure 5.1 shows
number 42 of his collection.

							1	5
			8	3				
						2		
	2	6				8		
				1				
	8							
1		5		4.				
			3			7	2	
9								

Figure 5.1

But what about Sudoku puzzles with fewer than seventeen entries and just one solution? Can they exist? This puzzle remains unsolved. As yet.

Postscript, December 2012:

On January 1, 2012, Gary McGuire from University College Dublin, together with Bastian Tugemann and Gilles Civario, have announced the results of a massive computer calculation—a computation that took more than 7.2 million core hours on a cluster of 320 computer nodes, performed in the twelve months from January to December 2012. According to this computation, there is no Sudoku puzzle with fewer than seventeen clues, since otherwise they would have found it. If the calculation is correct and complete (which has not yet been independently checked), then seventeen is the minimum number of clues. For me, the problem is still open: give me a convincing, nice, understandable, presentable argument why fewer than seventeen clues will not suffice. Even if the computation by McGuire and his group is correct, I have no clue.

3x + 1

Who has not tried, on receiving their first pocket calculator as a student, to see what happens when one presses the same sequence of keys over and over? Something like that must have been on the mind of Lothar Collatz (1910–1990) when he was a student in the 1930s. He did a simple calculation repeatedly (albeit without benefit of a pocket calculator) and stumbled on a pernicious little problem.

We start with an arbitrary positive integer, x, let's say 61. If the number is even, we divide by 2; if it is odd we multiply by 3 and add one. So from x we get either $\frac{x}{2}$ or $3x + 1$. We then treat this result as a new x and apply the same rule. Over and over again. Where does this lead to?

Let's try it. Start with $x = 1$; we then get 4, 2, 1 right back to where we started. When we start with $x = 3$, we get in order 3, 10, 5, 16, 8, 4, 2, 1—again ending at 1. Starting with 7 gives the sequence 7, 22, 11, 34, 17, 52, 26, 13, 40, 20, 10, 5, 16, 8, 4, 2, 1. Starting with the 61 suggested above, we get 61, 184, 92, 46, 23, 70, 35, 106, 53, 160, 80, 40, and from there we know how it goes. Would you like to try it with 27? If so, be prepared for a very long and loopy ride through the numbers, getting up to 9232. But after a while, you may have guessed it, we come back to 1. Does that mean that no matter what number we start with, we wind up, after some—possibly very long—sequence of numbers, with 1 and the short loop that continues from there ad infinitum? That is a question that has been repeatedly addressed in the past and has become festooned with a large number of names. One refers to it variously as the $3x + 1$ problem; as the Collatz, or Ulam, or Syracuse problem; or as the Kakutani Conjecture. Could some starting number yield an infinitely long sequence that never winds up in a loop? Or could some other number wind up in a loop different from 1, 4, 2, 1? Neither result can be excluded at this time.

Collatz unburdened himself of his considerations on this problem in his notebooks in the 1930s. In 1952 he infected his Hamburg colleague Helmut Hasse with the problem, and Hasse in turn took the infection to the Mathematics Department of Syracuse University in upstate New York. Stanisław Ulam was credited with the problem for a while. There was even a rumor that "the commies" had created the problem in order to divert mathematicians in the West and thus stunt serious mathematical research; the ruse was even presumed to have worked. It is possible that this rumor is due to an ironic cold-war comment; it may also have arisen because someone thought Ulam was a Russian (actually, he was a Polish Jew). Who knows? Another discoverer was Brian Thwaites, who happened upon the problem at 4 p.m. on Monday, the 21st of July, 1952, and promptly, if a little immodestly, called it Thwaites Conjecture.

Whomever one would like to crown as the discoverer of the $3x + 1$ problem, it has in the meantime been checked (with a computer) for numbers with up to eighteen digits by Tomás Oliveira e Silva. At least for that set of numbers the conjecture holds: any number with eighteen digits or fewer leads at some point back to 1. The cynical comment, "They all come down in the end," thus holds not only for airplanes but also for the zigzag, apparently random flight of the $3x + 1$ problem. But we don't have a proof for that yet. A proof was published in 1980 by M. Yamadas in the journal *Fibonacci Quarterly*, but it turned out not to be valid.

Is the $3x + 1$ problem important? In itself, not really. But it is precisely one of the recalcitrant small problems that shows us how complex iteration processes can be and how little we understand them. It is also connected with several difficult and important questions of number theory, such as the question of how close the powers of 2 and the powers of 3 can be, or (the same question expressed in more technical terms) how well numbers such as $\log_2 3$ can be approximated by fractions.

Should one work on such a problem? I mean, seriously? Well, of course, one doesn't have to, but one could, and it could be worthwhile. One can even work quite intensively and seriously on the $3x + 1$ problem, as a book by Professor Günther Wirsching, of the University of Eichstätt, showed. Several prizes have been offered for a solution of the problem, among them $50 by the geometer H.S.M. Coxeter in 1970, $500 by Paul Erdős, and £1000 by Brian Thwaites. Caution is indicated: at this point it is not at all clear what methods one might use to make the problem solvable.

Maybe Erdős was right when he said that mathematics is not yet ready for such problems. But that was twenty years ago.

The Perfect Monster

Already in antiquity, numbers were called "perfect" when they are equal to the sums of all their divisors. The smallest perfect number is 6:

$$6 = 1 + 2 + 3.$$

An almost religious aura surrounds the perfect numbers. The perfection of creation was explained by the fact that it took six days, and six is a perfect number. Or perhaps it was the other way around.

Most numbers are not perfect. If you try it yourself, you will find that in maybe around three quarters of the cases the sum of divisors is less than the number you started with. For 8, for example, the sum is $1 + 2 + 4 = 7$; for 9 it is only $1 + 3 = 4$, for 10 it is $1 + 2 + 5 = 8$. For another quarter of the cases the sum is too large: for 12 it is $1 + 2 + 3 + 4 + 6 = 16$ (this happens less frequently for small numbers, but becomes more common as the numbers get larger). Only very rarely do we find a perfect number where the sum matches the number exactly. After 6, the next larger perfect number is 28 = 1 + 2 + 4 + 7 + 14; we have to go to 496 for the next perfect number, and then all the way to 8128. And that's it, for the numbers below a million.

The history of perfect numbers spans millennia. It includes a memorable collaboration of Euclid with Euler.

In Euclid's *Elements* (Book IX, Proposition 36), we find (three centuries before Christ) the observation and proof that numbers of the form $2^{p-1}(2^p - 1)$ are perfect, provided that $2^p - 1$ is a prime number. We've already encountered prime numbers of the form $2^p - 1$ under the name "Mersenne primes." Clearly, Euclid knew about, and thought about, these numbers long before the monk Marin Mersenne (1588–1648) was able to put his name on them. That attribution is there-fore a little questionable. With Euclid's formula, the Mersenne primes 3, 7, 31, and 127 generate the perfect numbers 6 = 2 × 3, 28 = 4 × 7, 496 = 16 × 31, and 8128 = 64 × 127.

Leonhard Euler proved a sort of converse of Euclid's result: every even perfect number is generated by a Mersenne prime $(2^p - 1)$ with Euclid's recipe $2^{p-1}(2^p - 1)$. There are no other even perfect numbers.

At this time (December 2012), we know of forty-seven Mersenne primes, and thereby of course also forty-seven even perfect numbers. The largest known Mersenne prime, $2^{43,112,609} - 1$, has almost 13 million digits. The corresponding perfect number, $2^{43,112,608}(2^{43,112,609} - 1)$ is even more gigan-tic, with 26 million digits. We assume that there are infi-nitely many Mersenne primes; if that's correct, there are also infinitely many even perfect numbers. However, this has not been proven.

A greater mystery is the odd perfect numbers. Do any even exist? No one knows. No such number has ever been sighted—making them even more mysterious than the Loch Ness monster, for which there are at least alleged sightings as well as a fuzzy photo. The odd perfect numbers are, in a manner of speaking, the Nessies of mathematics. Perhaps there's even an infinite number of them, at least no one has yet excluded that possibility, but it's more likely that there are none.

But even though such an odd perfect number, a "Nessie-number" N_0, has never been seen, we do know quite a bit about it. I'll name only three such facts, each of them well established (proven): N_0 must have more than 300 digits and must thus be extremely large. It must have at least eight distinct prime factors—and at least eleven, if 3 is not one of the prime factors (by comparison, the perfect numbers generated by Euler's formula have just two prime factors: 2 and the Mersenne prime). And we know, thanks to a proof by Euler, that N_0 must be of the form power of some prime number times a square number.

So we actually know quite a bit about this number that no one has seen and that possibly (probably?) does not exist.

The Great Puzzles

Is there still anything new to be discovered in mathematics? Is there any progress? Are there breakthroughs? What are the great open problems? Which ones will be solved next? And why am I not working on any of them?

For the first three questions, the answer is a clear and definite "yes." The other three can't be answered with a "yes" or "no." Nonetheless, I'll give some answers here, even if that is naturally impossible.

Is there anything new to be discovered? Of course there is. After 500 years of mathematics as a separate discipline, we know a lot, but each solved problem raises at least two new ones. We have discovered fascinating structures, but the world of mathematics is *not* a sphere that can at some point be completely traversed, mapped, measured, photographed, and surveyed, like some giant Google Maps project, and then exhibited in a museum as a finished work. The world of mathematics is unbounded and infinitely large. Every day, new and fascinating objects or relationships or phenomena are discovered and described. Let me name just one: after decades of work, mathematicians and crystallographers determined that a crystal can have only three-fold, four-fold, or six-fold axes of rotational symmetry, never a five-fold axis. It is therefore impossible that a crystal structure could be rotated by one fifth of 360 degrees and made to coincide with the original structure. But then mathematicians discovered "aperiodic" structures that had a hidden five-fold rotational symmetry. The most famous of these was published by Roger Penrose in 1976; this "Penrose Tiling" of the plane uses two different quadrilaterals that Penrose called "kites" and "darts." A few years later, on April 8, 1982, Dan Shechtman, a crystallographer from Technion in Haifa, was amazed to find in his laboratory metallic crystals with a five-fold symmetry. This discovery in turn led to whole new fields of investigation, in both mathematics and in materials science, devoted to the theory and properties of quasicrystals. (It also led to the award of the 2011 Nobel Prize in Chemistry to Shechtman.) In June 2009 five-fold symmetric crystals were even found in nature: in a chunk of rock found in the Karyak mountains of the Kamchatka peninsula in far eastern Russia. So maybe not an accessible part of nature, but still....

Is there progress? Are there breakthroughs? Of course! Again and again! We are living in a golden age of mathematical research. In the last twenty years many important, longstanding problems of mathematics have been solved, partly on the basis of theoretical structures and insights whose foundations were laid by generations of mathematicians long after the problems were posed. Here are three of the best-known solutions.

1994

Fermat's Last Theorem says for any integer n larger than 2 there are no pairs of integers such that when each is raised to the nth power they add up to the nth power of another integer. (For $n = 2$, there are many such integers; they are just the Pythagorean triples, such as 3, 4, 5, and 5, 12, 13.) This proposition was written by Pierre de Fermat, a French jurist and mathematician, into the margin of a Greek book of mathematics with the remark that he had a really remarkable proof, but that it was too long to fit into the margin.

The solution to the problem—that is, the proof of Fermat's proposition—is, contrary to what Fermat said, very tough going and by no stretch of the imagination fit for the margins of a book. The proof was completed by Andrew Wiles in 1994 (in a second go-around, with the collaboration of Richard Taylor) and was published a year later. It is the brilliant pinnacle of a very tall theoretical edifice with very broad foundations. Fermat's "probably wonderful" proof, which must, after all, have used only the tools available to mathematics in the middle of the seventeenth century, has never been found. Presumably it does not exist, although the search continues, captivating mathematical amateurs who occasionally claim to have found solutions that turn out to be fictitious or simply erroneous. Fictitious mathematicians have also claimed to have proved the theorem, for example, in the science fiction novel *The Last Theorem* by Arthur Clarke and Frederik Pohl (2008), and in the detective novel *The Girl Who Played with Fire* (2009) by Stieg Larsson. The heroes of these books—the student Ranjit Subramanian from Sri Lanka and the Swedish computer savant Lisbeth Salander—find their real challenges in the theorem.

2002

The Catalan Conjecture states that there are no proper powers of integers that differ by just 1 except for $2^3 = 8$ and $3^2 = 9$. Eugéne Catalan posed this conjecture in 1844 in a letter to the editor of the *Journal für die Reine und Angewandte Mathematik* (the renowned *Crelle's Journal*, which has been published since 1826). It was finally proven in 2002 by Preda Mihăilescu of the University of Paderborn (and published also in *Crelle's Journal*, in 2004). Success—after 160 years!

With this proof, Mihăilescu provides a counterexample for the widely held belief (almost a cliché) that it is only young geniuses who contribute to mathematics. He was born in 1955 in Romania, fled at the age of eighteen to Switzerland where he worked his way from dishwasher to mathematics student, at which point he was earning a living as a programmer (in cryptography, which, as we have seen, involves a lot of number theory). He received his PhD from the ETH in Zurich in 1977 at age forty-two; two years later he was caught by the spell of the Catalan Conjecture. He was forty-seven by the time the solution came to him. In 2005 he was rewarded for his achievement with a professorship at the University of Göttingen.

2006

The Poincaré Conjecture and the even more far-reaching Geometrization Conjecture solve the problem of classifying the possible three-dimensional space forms ("manifolds") completely. The Poincaré Conjecture was universally recognized as one of the central problems of geometry. It was formulated in 1904 by the French mathematician Henri Poincaré and extended in 1980 by William Thurston to the much broader Geometrization Conjecture. In 2000 it was listed as one of the seven Millennium Prize Problems, for which the Clay Mathematics Institute offered a $1,000,000 prize. Its validity (as well as that of the Geometrization Conjecture) was proven by the Russian mathematician Grigori Perelman in 2002–2003, building on earlier work by Richard Hamilton. Several books have been published since 2006 that fill in the details of the proof's outline published by Perelman. At the International Congress of Mathematicians in 2006, the author of one of these books, John Morgan, declared the problem solved. At the same congress, Perelman was awarded the Fields Medal for his solution; he did not, however, accept the award and did not show up at the congress.

What are now the great open questions? Which of them will be the next to succumb to a solution? I will answer the first question with some science-fiction wishful thinking about answers for the second.

2015

On the 19th of May, Peter Sarnak, a sixty-one-year-old mathematics professor at Princeton, concludes his graduate course "Analytic Number Theory II" with a two-hour lecture in which he proves the Riemann Conjecture. The problem was formulated by Bernhard Riemann in 1859, almost as an aside in his only foray into number theory; it had been held to be the Holy Grail of number theory. David Hilbert listed it as one of the most important unsolved problems in mathematics in his famous lecture of 1900 at the International Congress of Mathematicians. In 2000, the Clay Foundation included it in the problems for whose solution they would give a prize of $1 million, turning it from a century problem into a millennium problem. The news of Sarnak's breakthrough reaches mathematicians worldwide within a few hours, even though only three of the students in the class were able to follow the lecture all the way through. The proof combines methods from number theory, algebra, and analysis with the physics of random systems. The basis for the rapid dissemination of the news is a set of lecture notes by one of the students, Isabella Scoli, only twenty-one years old, which in fact provide a substantial simplification of Sarnak's proof by showing how to avoid the use of difficult algebraic methods. After intensive review by seven referees, the proof appears in the summer of 2015 in the journal *Acta Mathematica*, published by the Mittag–Leffler Institute of Stockholm, as a joint paper by Sarnak and Scoli.

2015

Only three months after Sarnak's lecture, Nitin Saxena, a young Indian mathematician, announces the solution of another Millennium Problem: the $P \neq NP$ Theorem of complexity theory. The theorem concerns computational problems whose solutions, once found, can be checked effectively and efficiently, but for which there cannot be procedures to find the solutions in any reasonably rapid and efficient manner. As an example, Saxena points to the famous problem of the traveling salesman, whose task it is to determine the shortest path to visit a large number of cities without visiting any city twice. In an astonishing tour de force—the paper is only eleven pages long—Saxena provides an elegant combinatorial argument to show that the problem cannot, in fact, be resolved within the usual foundations of mathematics: it has neither a positive nor a negative solution; rather, the existence of a rapid method of computing solutions to problems like the traveling salesman problem depends on additional assumptions that one is free to make or not. Specifically, Saxena showed that whether $P = NP$ or $P \neq NP$ depends on whether one adopts the so-called continuum hypothesis—that is, whether one supposes that there exists a set containing more elements than the set of integers but not as many as the set of real numbers. Saxena's paper is a sensation and leads to intensive debates among mathematicians—but also in the general public. Saxena is invited to early morning and late night talk shows and appears on the covers of *Time* and *Newsweek* (or at least what passes for the cover for this now purely online publication). At the 2018 International Congress of Mathematicians in Rio de Janeiro, he is awarded the Nevanlinna Prize for Mathematical Aspects of Information Sciences.

2016–2019

Two years later, a fourth of the seven Millennium Problems is on the verge of a solution. A difficult 314-page paper by Maximilian Kreutzpaintner, a twenty-six-year-old Bavarian doctoral student at the Berlin Mathematical School, provides the techniques for solving the Navier–Stokes equations. These differential equations, formulated in the nineteenth century, remain incredibly important. They describe how fluids, such as air or water, flow. They are used to predict the weather and the effects of ocean currents as well as the flow of combustion gases through a jet engine; they are also used to optimize the performance of racing cars and airplanes by minimizing air drag. These solutions, however, give mathematicians headaches, because they are based on approximations and computational techniques rather than a firm theoretical foundation, which has remained incomplete until now. In particular, even the *existence* of solutions for three-dimensional flows (as, of course, all real flows are), in general, with all the possibilities of turbulence and vortices, has thus far not been clear at all. Because of this lack of theoretical foundation, the reliability of the computations that have thus far been used to predict, say, the aerodynamic performance of racing cars or airplane wings is somewhat suspect. Kreutzpaintner's manuscript presents a general solution of the Navier–Stokes equations, but it is also incredibly complex and difficult to read. It uses techniques that Perelman only sketched out in his solution of the Poincaré Conjecture and which have only been partially worked out. Many parts of the paper present only geometric and intuitive arguments instead of providing careful approximations. Reviewers find small errors and gaps in the argument, but the experts initially judge these not to be serious. Nonetheless the prepublication review, arranged by the journal *Forum of Mathematics Pi*, has stretched out over two years, when....

I will leave the further developments to your imagination (or to more professional prognosticators). Today, of course, we do not know what the future will bring. The only thing that is certain is that there will be progress, there will be breakthroughs, and some of these will lead to solutions of the seven million-dollar Millennium Problems.

And why don't I work on these "great" problems? Mathematics today is an incredibly large, rich, and multifarious field. There is so much to work on. I happen to be fascinated by discrete geometric structures and like working with these. Even in this field there are many unanswered questions, such as, how *complex* (in a strictly measurable sense) can a three-dimensional structure composed of triangular faces joined at edges and vertices really be? This problem suits me because I like geometry; because my personal tool kit of methods, ideas, interests, and technical knowledge is appropriate for the problem; and because I've accumulated some experience with problems in this area of mathematics and can thus hope to make a breakthrough. My colleagues in this field will be interested in the progress I can make, and we will all take joy when I, one of my students, or a colleague elsewhere in the world solves one of its problems. What we do, my colleagues, our students, and I, is interesting and worthwhile (in my opinion, of course, but also that of many others). But even if I do manage to solve one of my personal "great problems," there won't be any big headlines on page 1 of the newspapers, and maybe not even in the science section. But that's just fine by me.

Chapter 6

Where Mathematics Is Created

If one asks a mathematician about when and where some decisive idea came to life, one will rarely get a satisfactory answer. Mathematical ideas do not arise in a laboratory or other specific workplace, and only rarely do they come according to a plan or a well-laid session of intense thought. Instead, they arise at random points as little steps or larger jumps along a path that courses through the mathematician's life.

To make mathematics, one does not need much. Paper and pencil are the standard tools (I prefer graph paper, a soft pencil, and an eraser), the thinking is done with the head, and the checking of computations takes place on the laptop. The equipment in a mathematics laboratory is modest and also portable. Serge Lang tells a nice story about an engineering department that received a letter from the dean: "Why do you need so much money for equipment? Why can you not be more like mathematicians, who need only pencils, paper, and erasers; or like the philosophers, who need only pencil and paper?" Someone who wants to develop ideas as a mathematician needs to find time to think, needs to follow trains of thought, needs to have peace and quiet or diversions, and needs to concentrate intensely or relax and think freely. To develop ideas as a mathematician, one needs to be able to take the mind on a trip. There is no general recipe for success. Mathematics is highly variegated, and mathematicians are multifarious, even if they are often just stereotyped as gray-haired eccentrics with thick glasses. The places in which mathematics is done reflect the great variety of mathematical thinking styles. Let us therefore make an excursion into this diversity and approach, anecdotally, answers to the questions of where and how mathematicians work, and where and how mathematics arises.

At the Desk

"The mathematician is a mythological beast, half man, half chair," at least according to Simon Golin, a German management guru trained as a mathematician. And, of course, a lot of mathematics is created at a desk. Sometimes a decisive idea arises spontaneously in the course of a lengthy computation, or in the midst of a sequence of sketches, or while working out a set of examples at the desk. The desk here serves as a place of repose, of concentration, free from distractions.

Free from distractions? It is reported that Leonhard Euler was able to work and write effectively at his desk while his many children played around his legs and climbed around on his back. However, Euler, who was one of the most productive mathematicians of modern times, was apparently not easily distracted by anything. Even becoming blind in 1771 did not diminish his productivity, for almost half of his work was written after that time.

And what happened to the desk at which Euler worked? One can see the heavy, primitive-looking lab table at which Lise Meitner and Otto Hahn discovered nuclear fission at the Deutsches Museum in Munich. The desk at which Einstein worked in Princeton is still in the same office and is used by someone else now; Thomas Mann's study, with desk, can be admired in Zurich. But as far as I know, there is only one mathematical workplace that is preserved in a museum: that of Leopold Vietoris (1891–2002) in the "House of Mathematics" in Salzburg, Austria. I suspect that this will remain a singular example, since this desk has only limited value as an indication that mathematics arises at desks. Vietoris was an enthusiastic Alpinist even at an advanced age. When he was asked, at age ninety-eight, what his most valuable work was, he did not refer to his mathematical work, such as his dissertation of 1921, but rather wrote:

> In my opinion, my publications concerning the orientation on mountain climbers during periods of poor visibility (night and fog) have not received sufficient attention…. If these essays had been read and understood, then several Alpine accidents, with many fatalities, could have been avoided.

If nothing else, this shows that the mathematician Vietoris did not limit his imagination and experiences to the confines of his desk.

At the Coffee Machine

The legendary Hungarian mathematician Paul Erdős (1913–1996) provides many aspects and anecdotes for the creation of mathematical ideas. Erdős was a traveler, perpetually on the road, always a guest somewhere in the world where there were mathematicians. There was one thing that was constant, however: when Erdős arrived somewhere, he would settle himself on the sofa in the study or living room of his host and ask for a cup of coffee. Once that was in hand, he would lean back and say, "My brain is open." With a sofa and a coffee, it seems, the minimal conditions for a conversation about mathematics are fulfilled. And ideas can come then, too.

According to Erdős, a mathematician is a machine for transforming coffee into theorems. In my experience, however, there is little correlation between the quality of the coffee and the quality of the theorems. During my time as a graduate student at the MIT Mathematics Department in the 1980s, very bad coffee was turned into some excellent mathematics.

In addition to the coffee, the coffee machine is also an important component, serving as a meeting point. Both fellow students and professors can be approached while they are standing next to the machine and none will refuse the request, "May I ask you a question?" This was certainly the case when I was a student, and it still functions that way today. The lounge for my group at the Technical University of Berlin was called the "Molotov-Cocktail Depot": next to two rolling blackboards, there was a Swiss espresso machine that was highly useful for caffeine infusions—so useful that even confirmed tea-drinkers clustered around it. I recall that one Saturday afternoon (it was exceedingly quiet at the university, very few people were there) a conversation with my Austrian doctoral student Julian Pfeifle led to the decisive idea concerning "many triangulated 3-spheres," which led to a paper published a year later in the renowned *Mathematische Annalen*.

Erdős himself needed stimulants and sleeping aids in addition to caffeine. After a month of doing without the pills (due to a lost bet), Erdős declared, "It was a bad month for mathematics."

At the Café

To sit for hours in a café, thinking about a problem, scratching notes, formulas, and sketches into a notebook; drinking too much coffee; maybe even smoking; occasionally entering into discussions with other customers, some of them regulars like yourself—all that does not necessarily suggest productive work. But it can be. Myself, I've spent many days in a café, not only in the process of writing this book but also while trying to solve some problem in mathematics, sometimes success-fully, sometimes not.

The discovery that one can pursue mathematics in a café is not new. Several cafés and several café mathematicians have become important and famous. The Scottish Cafe in Lviv, for example, or the café Reichsrat in Vienna are places in which mathematics was created, developed, or announced.

The city of Lviv is today the seventh-largest city in the Ukraine; it was formerly in Poland and called Lwów; before that it was part of the Austro–Hungarian Empire and called Lemberg in German. In the 1930s there were the Café Roma and the Café Szkocka (the Scottish Café) on a small square only a hundred meters from the university. Who discovered the mathematical possibilities of these cafés? Perhaps it was Stefan Banach (1892–1945), an eccentric who loathed examinations and therefore never sat for one, but who was so brilliant and productive that he was finally awarded a doctorate anyway. Stanisław Ulam (1909–1984), who joined the discussions at the cafés as a student in Lwów, reported that Banach would spend entire days at the Café Roma, especially toward the end of the month, before the university's salary checks were issued. At some point Banach transferred his allegiance to the neighboring Café Szkocka, because of "more favorable credit arrangements," and his colleagues and students joined him there. The chemists and physicists remained at the Café Roma, while the Scottish Café henceforth became the residence of mathematicians. In his autobiography, *Adventures of a Mathematician,* Ulam tells of seventeen-hour-long discussions with Stefan Banach and Stanisław Mazur, discussions interrupted only by meals. These were serious work sessions in which problems were reviewed, results were formulated, and proofs were found.

The lasting fame of the Scottish Café, however, rests on a notebook maintained by the headwaiter in which unsolved problems were noted, often with a prize offered for the solution. The solution for Problem Number 6, for example, comes with a bottle of red wine offered by Mazur, who also offers two small beers for the solution of Problem Number 8. Both problems have yet to be solved. The notebook from the Scottish Café was published in 1981 as a book.

Café Reichsrat (at Reichsratsstrasse 13 in Vienna) has played a central role in the history of mathematics since at least August 26, 1930. On that date Kurt Gödel announced his Incompleteness Theorem. The event was noted stenographically in a diary entry by Rudolf Carnap:

> 6 – ½ 9, Café Reichsrat ... preparations for the trip to Königsberg ... Gödel's discovery: incompleteness of the systems of *Principia Mathematica* ... difficulties of the consistency proof

Gödel clearly presented his result first to friends and colleagues at the café. Happenstance? Certainly not. Kurt Gödel and his brother lived at the time in a rental apartment across the street from the mathematical seminar; the Café Reichsrat and many others were very close by. Students and faculty spent much time in the Café Josephinum, which was in the same building as the seminar, but also in the Café Arkadenhof just down the street, in the already mentioned Café Reichsrat, in the Café Central a few blocks away, and in the Café Herrenhof; several of these have survived and still serve excellent coffee. At the time these were quite smoky hangouts; today, in the interest of the waiters' and customers' health, they are smoke free.

Gödel's theorem is a major result of the century. It says that in the edifice of mathematical arguments there *must* exist theorems for which there are no counterexamples (and which are therefore true), for which, however, no formal proof can be provided based on the fundamental axioms of mathematics (and thus are not provable). Gödel's fame is based on this result, as well as a few others (he published very little, for he kept revising his papers, apparently never satisfied with them). When he was awarded an honorary doctorate by Harvard University in 1952, the citation named him as the "discoverer of the most significant mathematical truth of this century, incomprehensible to laymen, revolutionary for philosophers and logicians." At the time, Gödel wrote his mother in Vienna, "You should not understand this to mean that I was named as greatest mathematician of the century. The word 'significant' should rather be taken to mean: of the greatest general interest outside of mathematics." But then he writes further: "It was to be expected that my proof would be taken up by religion sooner or later. This is justified in a certain sense."

Is all that just history? Not at all. The basic principles of topology and functional analysis developed by the Polish mathematicians in Lwów are today essential foundations of mathematics. Without Gödel's theorems, modern logic and proof theory is inconceivable. And just as there are still literary coffee houses, there are also still mathematical coffee houses (I admit it: I'm a regular). The Scottish Café no longer exists, and the Café Reichsrat is today the Café Sluka, but there are still mathematics cafés all over the world, probably even more than in the 1930s. The mathematical economist Ariel Rubinstein of the University of Tel Aviv has published a List of Cafés on his web page. The list (like mathematics itself) is international and by now includes more than a hundred cities. Rubinstein calls it a worldwide travel guide for cafés "in which one can not only work but also think." As his contact information, he lists first the "Univ. of Tel Aviv Cafes" and then the Department of Economics.

In the Computer

Do we still need mathematicians to do mathematics? Couldn't we just get computers to do that? This question, in fact, has a long history. Since the 1950s, under the rubric "artificial intelligence," the ability of computers to "think" has repeatedly been announced with great fanfare. Among others, Herbert Simon (Nobel Prize for Economics, 1978) and Allen Newell have worked since 1957 on a computer program they call "General Problem Solver," which its authors describe as "a program that simulates human thought." However, for a long time there was a huge gap between the grandiose promises, like those emanating from MIT's Artificial Intelligence Laboratory where Marvin Minsky and Joseph Weizenbaum were active, and the decidedly modest feats that computers were actually able to achieve. Admittedly, the great Joseph Weizenbaum had already been issuing warnings about excessive faith in computer programs since the 1960s, and especially after the success of his psychotherapy computer program Eliza. And Gian-Carlo Rota writes:

> The computer is merely a tool for completing more quickly something that we can already do more slowly. All the arrogant promises of computer intelligence and expectations of paradisiacal possibilities should be turned down, otherwise the public will soon turn away in disgust. If that should happen, our civilization may not survive.

A computer cannot have ideas and therefore cannot do mathematics. However, there are discoveries that would not have been possible without a computer. I will show you a discovery made by one Roy D. North in the 1970s in Canada, based on results from a computer. It concerns the famous formula discovered by Euler in 1734 that we already discussed in the first chapter: if one adds up the reciprocals of all the squares of the integers—that is,

$$1+\frac{1}{4}+\frac{1}{9}+\frac{1}{16}+\frac{1}{25}+\frac{1}{36}+\frac{1}{49}+\frac{1}{64}+\frac{1}{81}+\frac{1}{100}+\frac{1}{121}+\frac{1}{144}+\frac{1}{169}+\frac{1}{196}+\frac{1}{225}+$$

one obtains, according to Euler, one sixth of π^2.

One can take this computation seriously and start to add. Of course, we won't finish because Euler's sum is an infinite series, so that there are infinitely many terms to be added, but we can start. The first partial sum is 1; the sum of the first two terms is 1.25; the first three fractions add up to:

$$1+\frac{1}{4}+\frac{1}{9} = 1.361\ 111\ 111\ 111\ 111\ 111\ 111\ 111\ 111\ 111\ 111\ 111\ 111\ 111\ 111\ 111$$

To this, we have to add $\frac{1}{16} = 0.0625$, and then $\frac{1}{25} = 0.04$, and so forth. The more terms we add together, the closer we get to Euler's result, which one can write as:

$$1+\frac{1}{4}+\frac{1}{9}+\frac{1}{16}+\frac{1}{25}+\frac{1}{36}+\frac{1}{49}+\frac{1}{64}+\frac{1}{81}+\frac{1}{100}+\frac{1}{121}+\frac{1}{144}+\frac{1}{169}+\frac{1}{196}+\frac{1}{225}+\frac{1}{256}$$

$$= 1.644\ 934\ 066\ 848\ 226\ 436\ 472\ 415\ 166\ 646\ 025\ 189\ 218\ 949\ 901\ 206\ 798\ 437\ 735$$

$$558\ 229\ 370\ 007\ 470\ 403\ 200\ 873\ 833\ 628\ 900\ 619\ 758\ 705\ 304\ 004\ 318\ 962\ 337$$

If one adds up only a finite number of terms, one of course does not get $\pi^2/6$ but something a little less than that. So, for example, the sum of the first one million fractions—added by a computer, of course, carefully programmed to compute precisely—produces the partial sum:

$$1+\frac{1}{4}+\frac{1}{9}+\frac{1}{16}+\frac{1}{25}+\frac{1}{36}+\frac{1}{49}+\frac{1}{64}+\frac{1}{81}+\cdots+\frac{1}{1,000,000,000,000}$$

$$= 1.644\ 933\ 066\ 848\ 726\ 436\ 305\ 748\ 499\ 979\ 391\ 855\ 885\ 616\ 544\ 063\ 941\ 294\ 911$$

$$748\ 705\ 560\ 407\ 903\ 303\ 634\ 027\ 380\ 082\ 445\ 906\ 638\ 492\ 190\ 883\ 355\ 611\ 907\ 104\ 4$$

As you can see, the first five digits are already correct. The sixth digit is not, but this should not surprise us: it is well known that in adding up fractions, the partial sums approach the correct result only slowly. But what is surprising is that the seventh, eighth, ninth, tenth, eleventh, and twelfth digits are exactly correct, while the next is not. Incorrect digits occur, it seems, at roughly the sixth, twelfth, eighteenth, etc., positions, and the size of the errors is also quite systematic—involving the Bernoulli numbers, $0, -\frac{1}{2}, \frac{1}{6}, 0, -\frac{1}{30}, 0, \frac{1}{42}, \cdots$ in an important way. These insights are not easily seen on a casual inspection, even for masterly calculators like Euler, Gauss, and Riemann; they could only have been found with a modern computer using highly developed mathematics in the form of sophisticated software.

But after that, formulating the result mathematically and then proving it can only be done with pencil and paper. In the case of the approximations of Euler's formula, that was first achieved by the brothers Jonathan and Peter Borwein together with Karl Dilcher many years after its discovery by Roy D. North in his computer printouts.

In Bed

Four weeks before his nineteenth birthday, Carl Friedrich Gauss made an important discovery—namely, that one can construct a regular seventeen-sided polygon with a ruler and compass, and nothing more. This is a remarkable discovery: it is the first time in something like 2000 years that such a construction was discovered. Euclid had known how to construct equilateral triangles, squares, and regular hexagons—and we all learned how to do that in our geometry classes in school. Euclid also knew how to construct a regular pentagon, though that's a little trickier (the method is described in the *Elements*). He did not know of a way to construct a regular heptagon, and Gauss proved that it is in fact impossible. Regular octagons are again possible, and so forth, until seventeen. Euclid found no way to construct the figure, and no one after him either, until the young Gauss was lying awake in bed one evening. He described the discovery in a letter:

> The story of that discovery has never been mentioned by me in public, but I can describe it to you precisely. The day was the 29th of March 1796, and chance had no part in it.... On the morning of the said day during a vacation trip in Braunschweig (before I had risen from my bed), thinking strenuously about the relationship of all the roots among each other according to arithmetic criteria, I succeeded in being able to view these relationships most clearly, so that I could see the special application to the 17-sided polygon, and I could make the numerical confirmation of the result on the spot.

And I suppose that "on the spot" means still in bed.

On the other hand, the story that Gauss jumped out of bed on his wedding night in order to make note of some mathematical formula, which is told by Daniel Kehlman in his satirical novel *Measuring the World* (2006), is probably not true, and we would not want to wish that for him—and even less for his wife.

There are two varieties of "in bed." One is the "lying awake at night, not being able to sleep, and then finally having a decisive idea" variety or the "waking up early and lying awake with thoughts going round in one's head" variety (this may be what happened to Gauss in 1796). The other is the inspiration while one is asleep, a dream perhaps, that needs a little, or sometimes a lot, of interpretation. G. H. Hardy, in his memorial lectures for Srinivasa Ramanujan (1887–1920), reports that

> Ramanujan always said that the goddess Namakkal gave him formulas in his dreams. It is noteworthy that often, immediately upon rising, he made notes

of formulas and also quickly verified them, although he could not always provide a rigorous proof.

The divine inspiration of the dreams did have an unfortunate catch, however: the formulas were indeed noteworthy, but not always correct. Ramanujan, who was for Hardy "the most romantic figure in the history of modern mathematics," was a gifted genius who worked as a clerk at the port of Madras; he had not completed his formal education and had no training in higher mathematics; and he found a large number of wonderful formulas, many of which he could not prove himself—and some of them have proven hard nuts to crack. Mathematicians, starting with G. H. Hardy and John Littlewood, who were his hosts in England, and subsequent generations up until the most recent times, have worked on proving Ramanujan's formulas and interpreting his "mock theta functions," including Ken Ono of Atlanta and Kathrin Bringmann of Cologne, the recipient of the Alfried-Krupp Prize for young professors in 2009. Nothing comes for free.

In Church

Divine inspiration is probably difficult to confirm in a mathematical discovery. But why should it not be possible for the festive atmosphere of a service in the Vatican (complete with intoxicating fumes from incense-burning censers) to lead to ideas?

In his obituary for Johann Peter Gustav Lejeune Dirichlet (1805–1859), Ernst Eduard Kummer writes

> The clarity and certainty of his thinking and the unusual strength of his memory through which he was able at all times to keep in mind all that he had explored or thought made the use of a pen while working almost entirely superfluous. He did not

need any particular calm or leisure for this but could pursue his deep speculations during walks or travels, musical diversions, or indeed in all situations where he was not himself called upon to speak or act, with the same success as at his desk. As an example I can cite a difficult problem in number theory with which he had wrestled in vain for some time, whose solution came to him in the Sistine Chapel in Rome while listening to the Easter music that was traditionally performed there.

Divine inspiration? And if so, how does one acknowledge it appropriately?

One often finds expressions of gratitude among the acknowledgments in professional mathematics articles. It is generally considered good form to thank colleagues for suggestions, discussions, and ideas that contributed to the research being described. Less frequent are expressions of thanks to perhaps temporarily neglected friends or spouses—but then, these are more personal matters that may not need to be part of the published record of science. On the other hand, it has become exceedingly rare to express thanks for divine inspiration. Indeed, such expressions can become a little awkward, particularly if coauthors are not fully in agreement:

> The first author would like to acknowledge and thank Jesus Christ, through whom all things were created, for the encouragement, inspiration, and occasional hint, that were necessary to complete this article. The second author, however, specifically disclaims this acknowledgement.

This acknowledgment can be found in the renowned journal *Mathematische Zeitschrift*, volume 247 (2004).

In Captivity

The French mathematician Jean-Victor Poncelet (1788–1867) developed the field of projective geometry while he was a prisoner of war in Russia. He was a lieutenant and engineer in the French army and participated in Napoleon's invasion of Russia. After the battle of Smolensk in 1812, he was left for dead (his horse had, after all, been shot out from under him) and was thus taken prisoner by the Russians after the battle. He was marched for four months through the Russian winter some 1500 kilometers to the east, ending up in Saratov on the Volga, where he remained for two years before being able to return to France. In the prisoner of war camp, without any books or access to a library, Poncelet developed the foundations of modern projective geometry, which became his magnum opus, *Traité des proprietés projectives des figures*, published in 1822. (The second edition of 1862 included his reminiscences in the preface.)

The Austrian Edmund Helly (1884–1943), from Vienna, was drafted into the army in 1915. In that same year, his lung pierced by a gunshot, he was captured by the Russians. Held in a Siberian camp near Nikolsk-Ussuriysky, he finally returned to Vienna in 1920 via Japan and Egypt. In Vienna he was able in 1921 to get a position at the university with his paper "Concerning Equations with an Infinite Number of Unknowns," which he had written by hand during his captivity. Professor Hans Hahn, who had recently been appointed to the University of Vienna and whose name every mathematician knows because of the Hahn–Banach Theorem, honored Helly by saying that he "delved considerably more deeply into the theory of systems equations with infinitely many unknowns than any of his predecessors."

Leopold Vietoris (1891–2002), also an Austrian, was born in Radkersburg in Styria. In 1914 he volunteered for the army and was wounded in the fall of 1914; after his recovery he was assigned as a mountain guide to the southern front and was captured by the Italians just before the end of the war in 1918. He was apparently treated sufficiently well that he was able to finish his dissertation on "Continuous Sets," which he submitted to the University of Vienna after his release. The wartime injury and captivity seem not to have damaged his health too seriously, since he was for many years the oldest living Austrian; he died in 2002 shortly before his 111th birthday. A story is told of an excursion in the 1990s of the Austrian Academy of Sciences to the baroque monastery at Melk on the Danube during which Vietoris mentioned, by the way, that he had gone to school there "a century ago."

Jean Leray (1906–1998) developed his deepest and most important insights, which became fundamental contributions to modern algebraic topology, specifically "spectral sequences" and the theory of "sheaves," in an Austrian prisoner of war camp for French officers. Leray was interned in Edelbach for nearly five years, from July 1940 until the end of the war in May 1945.

The French number theorist André Weil (1906–1998) was held in a detention camp near Rouen as a conscientious objector. His colleague Élie Cartan wrote him, probably intending to cheer him up, "We do not all have the good fortune to be able to work in peace and as undisturbed as you." In the end, Weil was apparently able to use his time productively. He wrote to his wife:

> Actually, since I saw you, my arithmetico-algebraic research has got off to a good start. I have found some interesting things—to the point where I'm hoping to have some more time here to finish in peace and quiet what I've started. I'm beginning to think that nothing is more conducive to the abstract sciences than prison. My Hindu friend Vij often used to say that if he spent six months or a year in prison he would most certainly be able to prove the Riemann

hypothesis. This may have been true, but he never got the chance.

And further, quite cheerfully:

> My mathematics work is proceeding beyond my wildest hopes, and I am even a bit worried—if it's only in prison that I work so well, will I have to arrange to spend two or three months locked up every year? In the meantime, I am contemplating writing a report to the proper authorities, as follows: "To the Director of Scientific Research: Having recently been in a position to discover through personal experience the considerable advantages afforded to pure and disinterested research by a stay in the establishments of the Penitentiary System, I take the liberty of, etc. etc."

Probably not completely serious. Gallows humor seems more likely.

In an Attic Room in Princeton

The proof of Fermat's Last Theorem—that is, that the equation $x^n + y^n = z^n$ has no integer solutions for x, y, and z for any n larger than 2—which Andrew Wiles achieved in the 1990s, ranks among the more dramatic stories of modern science. That someone could withdraw himself to an attic room for seven years to solve one of the great mathematical problems, that his solution is then heralded with a flash-photo storm of paparazzi, that the solution is then found to have a fatal flaw, that our hero should then withdraw into his attic again, and while he cannot correct the flaw he finds a way to circumnavigate it with a new idea. Is this not comparable to some of the heroic deeds of antiquity?

One can wonder why someone has not dramatized this story—perhaps not as a Homeric epic in iambic pentameter or as a drama for a Greek amphitheater, but rather as something modern like a play or Broadway musical. An excellent project, one would think, for, say, a British science writer. That it's highly unlikely to happen is, perhaps, a sign of the times.

Andrew Wiles has reported on *how* and *where* the decisive ideas arose:

> Much of the time I would sit writing at my desk, but sometimes I could reduce the problem to something very specific—there's a clue, something that strikes me as strange, something just below the paper, which I can't quite put my finger on. If there was one particular thing buzzing in my mind then I didn't need anything to write with or any desk to work at, so instead I would go for a walk down by the lake. When I'm walking I find I can concentrate my mind on one very particular aspect of a problem, focusing on it completely. I'd always have a pencil and paper ready, so if I had an idea I could sit down at a bench and start scribbling away.
>
> I was sitting at my desk one Monday morning, 19 September, examining the Kolyvagin–Flach method. It wasn't that I believed I could make it work, but I thought that at least I could explain why it didn't work. I thought I was clutching at straws, but I wanted to reassure myself. Suddenly, totally unexpectedly, I had this incredible revelation. I realized that, although the Kolyvagin–Flach method wasn't working completely, it was all I needed to make my original Iwasawa theory work. I realized that I had enough from the Kolyvagin–Flach method to make my original approach to the problem from three years earlier work. So out of the ashes of

Kolyvagin–Flach seemed to rise the true answer to the problem.

It was so indescribably beautiful; it was so simple and elegant. I couldn't understand how I'd missed it and I just stared at it in disbelief for twenty minutes. Then during the day I walked around the department, and I'd keep coming back to my desk looking to see if it was still there. I couldn't contain myself, I was so excited. It was the most important moment of my working life. Nothing I ever do again will mean as much.

In remembering this moment, Andrew Wiles is moved to tears—tears that we, as well as future generations, can all witness, for they are preserved in the BBC documentary *Fermat's Last Theorem* by Simon Singh and John Lynch. After the first broadcast of the program on January 26, 1996, the *Guardian* said, mathematics is the new rock and roll.

On a Beach

Mathematicians can of course work at the beach, sometimes with legendary results. Stephen Smale describes the circumstances for his work in 1960 in Rio de Janeiro:

In a typical afternoon I would take a bus to IMPA and soon be discussing topology with Elon, dynamics with Mauricio or be browsing in the library. Mathematics research typically doesn't require much, the most important ingredients being a pad of paper and a ballpoint pen. In addition, some kind of library resources and colleagues to query are helpful. I was satisfied.

Especially enjoyable were the times spent on the beach. My work was mostly scribbling down ideas and trying to see how arguments could be put together. Also I would sketch crude diagrams of geometric objects flowing through space, and try to link the pictures with formal deductions. Deeply involved in this kind of thinking and writing on a pad of paper, the distractions of the beach didn't bother me. Moreover, one could take time off from the research to swim.

Smale became famous for, among other things, his work in Rio, which included his proof of a many-dimensional version of the Poincaré Conjecture as well as important insights into the theory of dynamical systems. However, his claim that he did some of his best work on the beaches of Rio brought him much grief. When he was under attack for his outspoken support of the protests against the Vietnam War, the science advisor to the president accused him of "wasting taxpayer money" on the beaches of Rio.

That was in the 1970s. However, even nowadays a workday at the shore can lead to curious controversies. An editor at SIAM, the Society for Industrial and Applied Mathematics, wanted to remove the following passage from a Festschrift for the mathematician Manfred Padberg, because she found it discriminated against women:

> One may bump into Manfred here, there and everywhere, Berlin, Bonn. Lausanne, New York, Tampa, Hawaii, Grenoble, Paris, but do not interpret his work on the Traveling Salesman Problem in the context of his own peregrinations. If you meet him on the beach in Saint-Tropez, he will be very likely working on a portable, without a look to the sea or to a group of attractive ladies! My personal opinion

is that Manfred Padberg is the perfect specimen of a new type of man, one who prefers spending his time in front of a computer. Maybe after Homo Erectus, Neanderthals, Cro-Magnons, and Homo Sapiens, we are confronting a new breed of *Homo mathematicus?*

In a Paradise with a Library

For many mathematicians the perfect work environment is in a place like the Mathematical Research Institute Oberwolfach. An American mathematician is said to have greeted those who arrived after him with the words, "Welcome to the mathematicians' paradise." This research and conference center is in an isolated part of the Black Forest in southwestern Germany. This is not a failing, but an advantage. The center has good food, quiet repose, an excellent library, large blackboards, a copy machine, computers, several espresso machines, a billiard table, a ping-pong table, a music room, a large wine cellar, long paths in the woods, and colleagues from all over the world. Visitors can use any one of these components as they see fit, excessively, moderately, or not at all; however, each one of them has played an important role in the creation of ideas—provably!

One example is the red ping-pong paddle from Oberwolfach on which Gerhard Frey (of Essen) enthusiastically explained to Günter Harder (of Bonn) his idea that the Fermat problem has some connection to truly strange elliptical curves—which, as it turns out, is actually the case and was the key to Wiles's proof. That paddle is said to be still in existence somewhere.

The most important treasure of the institute is, however, its excellent library. Many of the best new mathematical ideas are connections between old ideas. To come up with such new ideas, one of course has to know the old ones. The established old ideas can be found in the library; the established current ideas are known to the colleagues. So all one has to do is put the things together properly. And indeed, that does happen again and again. A tactical search in the library can sometimes bring the necessary resources to light: a lemma by Gauss from 1801 or a formula by Euler from 1748. Unlike publications in, say, biomedical or computer sciences, which become out of date rapidly and are then no longer read because knowledge in these fields is constantly replaced by newer knowledge, the insights of mathematicians retain their value and their truth.

And so every week almost fifty mathematicians from all over the world congregate in Oberwolfach, present their newest considerations, discuss their ideas at the billiard table or over a glass of red wine, and poke around the library.

Secluded in the Black Forest: that is the intention. Whereas radiologists meet each summer for a large conference in Berlin (well funded by big companies looking for customers for their imaging tools and products), and computer scientists meet annually in large hotels in Boston, Hawaii, or Singapore, mathematicians meet in small groups in Oberwolfach, each week a different group. And because this model is so successful and fits so well to mathematicians, it is no longer unique. There are now similar centers for mathematical meetings and workshops at Luminy near Marseilles (in an oak forest, within running distance to the Mediterranean), at Djursholm near Stockholm (in the charming art-nouveau villa of the mathematician Gösta Mittag-Leffler), in Berkeley, near San Francisco (high up in the Berkeley Hills, with views of the bay and of sunsets behind the Golden Gate bridge), in Bedłewo, near Posnań (a former Polish nobleman's estate), and in Banff, high up in the Canadian Rockies.

Knowledge in the ArXiv

The Internet is visibly changing the way science is done, including the science of mathematics. This is especially true of the role of libraries and technical journals. These days, mathematicians who think they have discovered something substantial discuss it with their colleagues, write it up, and then send it first to ArXiv.org, a science library on the Internet, which opened its virtual doors to mathematics on December 1, 1997. By 2008 it was accepting 14,370 technical papers in mathematics per year.

The ArXiv site publishes preprints, that is, preliminary versions of articles; before ArXiv, scientists would mail out preprints to colleagues interested in the subject and hope to stimulate discussions or comments; the ArXiv saves mailing expense and widens the distribution. Nowadays, authors will send an article to the ArXiv and will then send it to a technical journal; the journal then reviews the article, usually by sending it to referees who remain anonymous, which can easily take six to twelve months. Even if the reviewers are in favor of publication, they often suggest corrections or improvements, which the authors then have to take care of, before the article can be published in the journal with an imprimatur "reviewed and found correct and interesting" for the readers of the journal. After all that, the final printed version may follow the preliminary ArXiv version by one or several years—and will be available only to those whose library subscribes (often at great expense) to the journal, while the ArXiv version is available everywhere for free.

Grigori Perelman, whom we encountered in the previous chapter, saved himself the lengthy peer-review process. This ingenious (but also highly eccentric) mathematician from St. Petersburg presented his solution of the Poincaré Conjecture in three articles simply deposited on ArXiv: the first, with thirty-nine pages, on November 11, 2002; the last, with only seven pages, on July 17, 2003.

The Poincaré Conjecture, one of the great problems of the twentieth century, was formulated early in the century; it withstood attack from many different directions with many profound tools for decades. Generalizations to dimensions greater than three were proved by Stephen Smale ("on the beaches of Rio") and by Michael Freedman (on the beaches of San Diego?), both of whom received Fields Medals for their efforts. But the original conjecture, which refers to the structure of three-dimensional space, remained open.

The Clay Foundation named the Poincaré Conjecture as a Millennium Problem and set a prize of a million dollars for its solution. Suggested solutions, both by professionals and by amateurs, became frequent. Someone bold enough to publish a solution on ArXiv will also have to live with the fact that the "solution" will remain posted on ArXiv for all to see, even if the solution is ultimately found to be wanting. It should also be noted that the Clay Foundation will pay out its million only for a publication in a peer-reviewed journal, not for a preprint. So even a proper solution will have to wait a year or two after the initial ArXiv submission for a payout.

In the case of Perelman, the ArXiv submission was taken seriously. He had already given a couple of well-regarded talks on the topic, one in Berlin and others at MIT and Princeton, at which he answered critical questions to the satisfaction of experts in the audience. But then he withdrew. Not one word more, no details, no explanations, no supplements. And also no submission to a technical journal. The experts of differential geometry were thus left with just the three preprints, which turned out to be extraordinarily dense and difficult sketches of a proof, with many details remaining to be fleshed out, but without substantial errors or gaps. In the end, it took three years before some completely worked-out proofs became available—in time for the International Congress of Mathematicians in Madrid in August 2006, where the official announcement could be made, "The proof is correct, the Poincaré problem is solved."

Perelman was apparently not impressed. He did not accept the Fields Medal that was offered to him at the congress. (A place had been reserved for him in the front row of the auditorium where the awards were presented, just in case he did decide to show up. Ultimately I wound up sitting in that place, but that's another story.) In the end, Perelman also did not accept the Clay Foundation's million dollars, when he was finally offered them in 2010. Perelman is considered a genius, but also, to put it politely, "difficult." He lives well, apparently, isolated in the forest around St. Petersburg and has contact only with his mother; occasionally he goes to the opera. I don't know that for sure. And neither does almost everyone else.

Research in the Internet?

Where will mathematics be done in the future? Well, of course, some will still happen at desks, on computers, in libraries, on beaches, at blackboards, in coffee houses, in beds, and at coffee makers. Maybe in the Internet?

Tim Gowers is a brilliant colleague in Cambridge, England. His brilliance is confirmed and recognized: he received a Fields Medal at the International Congress of Mathematicians in Berlin in 1998. In addition to his particular research contributions, he has also made important and notable contributions to the public perception of the science of mathematics, in particular with a thin little red book *Mathematics. A Very Short Introduction*, and with a large, fat, black tome, *Mathematics. A Very Long Introduction*.

On January 27, 2009, Gowers started a discussion on "massive collaboration" on his highly regarded blog. He asked, how large a group of mathematicians could work together to solve a problem. By a "large group," Gowers did not mean just "more than four" but "hundreds"! Such a collaboration would need an infrastructure—which already existed on the Internet, in the form of discussion forums, wikis, and other tools. These should be used, Gowers said.

The suggestion is attractive. Different mathematicians have different work styles that could, in the right environment, complement each other. Some mathematicians would generate ideas, others would provide deep knowledge of the technical literature, still others would connect disparate parts. A large group would generate many ideas very quickly—and with luck some of these ideas would come out of entirely different directions but would fit together perfectly to provide a solution. Gowers expressed the hope that no one mathematician participating in such a collaboration would have to work very hard, but that the whole Internet collaboration would work as a sort of super-mathematician, whose brain was the sum of all the brains of the individual participating mathematicians.

But is that realistic? To provide an example of the possibilities, Gowers posed a problem, a specific, unsolved problem in combinatorics, and invited collaboration. Six weeks later, and after more than a thousand entries and comments on the blog, the problem was solved. It then took another two months to write it up properly and submit it to ArXiv under the pseudonym "D.H.J. Polymath." So there is at least one successful, large-scale collaboration. But what if the success was just due to the fact that the collaboration included Tim Gowers and Terry Tao, two of the best mathematicians currently active? Would a second or third Internet collaboration work equally well? Is this a useful model for the future of mathematical research? I don't know.

Chapter 7

The Book of Proofs

My favorite book does not exist—at least, not as far as I know.

Paul Erdős often talked about the Book of Proofs, the Book into which God had written all the perfect proofs of mathematical theorems. Of course, since the Book is in God's hands, it is not available to us. But, according to Erdős, when God means well with us, he lets us glance at the Book for a moment. And that is one of the greatest pleasures for a mathematician: to see—or, even better, to find—a *perfect* proof. Mathematicians, said Erdős, don't really have to believe in God, but in the Book, yes.

I won't say any more about this, as I don't want any grief. When I was interviewed for the newspaper *Die Zeit*, for example, I made some comments about the Book; when the story was printed, it made it seem as if God did not let us look into the Book out of malice. Soon after, I received a long handwritten letter from a seventy-six-year-old reader who was a Jehovah's Witness, in which the alleged accusation of maliciousness was thoroughly refuted with numerous biblical citations. So I won't say any more about this.

But the proofs. Nothing works in mathematics without proofs. Proofs are the heart and mind of mathematics. They are the pillars upon which the structure of mathematics rests. And when we write about mathematics, we can, of course, not leave the proofs outside by the door. So this chapter is about the proofs.

About Proofs

Mathematics is a game with many questions and few answers. It is a game with an infinity of propositions but only a finite number of proofs. Proofs are the way we certify what we know for certain, and it is therefore worthwhile to look at just what it means to be a proof.

Here are five questions and five answers on the topic.

1. Who Invented Proofs?

The oldest written record that we have of the art of constructing formal proofs in mathematics is Euclid's *Elements*, written in 325 BCE, in which he summarizes the entire mathematical knowledge of the time and in which formal proofs are included.

The "Greatest Hits" of proofs from the *Elements*, proofs that are so clear and elegant that one can practically sing them or tell them as bedtime stories, are proofs concerning number theory. In the *Elements* we find the oldest and most classic of all proofs, the proof that there must be an infinite number of primes. For this you first convince yourself that every number is either itself a prime number or it can be written as a product of prime numbers. The proof then follows in essentially a single line: consider any finite set of primes; multiply them all together and call the result P; then add 1 to get $P + 1$. Now, either $P + 1$ is itself prime and is, of course, not in the original set, or it can be factored into a product of primes none of which can be in the original set because all the primes in that set are factors of P and thus can't be factors of $P + 1$. So no finite set can contain all the prime numbers. QED.

2. Why Do We Need Proofs?

That question is posed by many, not just by the engineering students in our required (and loathed) courses of "mathematics for engineers." The answer we give the engineers is that we need to be absolutely certain that the calculations you make will completely and reliably ensure that the bridges and skyscrapers you design based on those calculations will not fail or collapse.

Mathematics itself is a gigantic structure with a relatively narrow foundation. And it is not just applications in engineering but all the natural sciences and the structure of mathematics itself that depend on the reliability of mathematical reasoning. In mathematics, knowledge is certain only if it is supported by a proof, and not otherwise. Of course, there is a considerable amount of belief and supposition in mathematics, and much more of that than there is of the proven and certified—one can experiment, make sample computations, argue from experience, develop intuitions, all of which are useful. But the only thing that counts in the end is what has been proven.

3. Why Do We Need So Many Proofs?

Actually, a single solid and complete proof will suffice to ensure that a mathematical result is correct. A multiplicity of proofs does not provide additional levels of correctness or, consequently, of security, in establishing a result. (This is in contrast to the way proofs work in legal proceedings.)

Nonetheless, many of the correct propositions of mathematics have been proven several times over. For example, when he was seventeen, Paul Erdős was proud to know thirty-seven proofs of the Pythagorean Theorem. (I don't know if that list included the proof discovered by then-congressman and later president James A. Garfield in 1876, allegedly during a session of Congress.) A book by Franz Lemmermeyer, published in 2000, collected no fewer than 196 proofs of the law of quadratic reciprocity, first proven by Gauss in 1801; Gauss himself subsequently found seven further proofs. This is not just an academic exercise and collection mania. There is much to be gained, since each truly novel proof provides new insights and new ideas, and then also perhaps a key to open up extensions and generalizations of the result.

Correspondingly, there is also no one proof that is clearly the best or the most beautiful proof. Beauty is, after all, in the eye of the beholder, and that's true in mathematics also. The shortest proof may not be the clearest, and an elegant proof may not be the most general. While there are several proofs of the Prime Number Theorem (including one by Erdős), the most powerful proof has not been found by a long shot—it would at least have to include a proof of the Riemann Hypothesis. Some proof might be better algorithmically, that is, it might help us find prime numbers, while another might be less complex. But the *perfect* proof remains in the Book, as yet unseen.

4. What Does a Proof Look Like?

The image of a proof comprising a "chain of evidence" suggests that in a proof one argument follows another to form a connection without gaps between "what we know" and "what we want to demonstrate." What is correct about the image is that a chain fails if just one link fails, and the same is true of a proof: if any one of the intermediate steps is incorrect (or is not properly founded on what is already known), then the whole proof fails, and we know no more about the reliability of the conclusion than we did without the alleged proof. In other ways, however, the image of a chain is quite misleading. It suggests that a proof consists of a large number of links, each of a size and import just like the others, and each following the other in a linear arrangement. It suggests that at each step we should know what the next step should be and in which direction it will take us. If that were actually the case, mathematics would be a very dull enterprise. But it isn't.

Proving things is a creative, experimental, unplanned, artistic, and athletic endeavor. You can think of it like building a bridge in a dense fog: There is a goal (the other shore, the theorem); there are several possible anchor points on this shore (what is known), some of them perhaps already extending into the water a bit. Usually there is no clear plan, though there may be intermediate points (an island, whose position is approximately known, or a pier that someone has previously pounded in and which may still be able to support some weight). And then there are the construction materials, planks, reinforcing rods, cables, safety nets. Perhaps there is a lookout point from which the things can be seen more clearly. Of course there will be detours, sometimes tunnels. And, fortunately, there are teams of collaborators, for who wants to build a long bridge all alone?

The great achievements in mathematics of the last twenty years—such as Andrew Wiles's proof of Fermat's Last Theorem or Grigori Perelman's proof of the Poincaré Conjecture— are anything but sequences of small steps chained together. Think, rather, of towers of theories built upon other theories forming a complex interconnected edifice; the conclusions one reaches at the top, along with deep intuitions, then suggest what might be the next steps or places where one can anchor a new structure. And occasionally someone like Wiles or Perelman comes along and adds a whole new tower, building on parts of the structure that have been worked on for centuries. The proof of Fermat's Theorem can be said to have started with Gerhard Frey's ingenious insight that the proof might have something to do with elliptic curves, which led to the entire theory of elliptic curves that has been developed since the early nineteenth century (starting with Niels Henrik Abel in Oslo and Karl Weierstrass in Berlin). These are masterpieces made by giants standing on the shoulders of giants.

5. Can One, in the End, Find a Short and Simple Proof for Everything?

No.

Consider Fermat's Last Theorem. That story, as is well known, starts with a comment by Fermat in the margin of a book, saying that the equation $x^n + y^n = z^n$ has no positive integer solutions for $n > 2$, and adding that he has found an ingenious proof for this, but that the proof is too long to fit into the margin. It has concluded (so far) with Wiles's famous proof, published in the *Annals of Mathematics* in 1995, which starts with an abstract that does not summarize the proof but reproduces (in the original Latin!) Fermat's marginal comment, and then proceeds with 109 pages of incredibly difficult mathematics that can't be understood without a pile of technical references and journal articles.

Is this the proof Fermat had in mind?
Clearly not.

Does the proof that Fermat had in mind exist?
Probably not.

Why not?
Because hundreds of clever mathematicians (women, such as Sophie Germain, as well as men) have wracked their brains on the problem. Not only have they not found a simple proof, but they have also been able to argue convincingly that there is no proof that uses only the mathematical methods Fermat had at his disposal (which includes all of elementary number theory, partitions, induction, infinite descent). However, that has not yet been proven.

And what if I've found a nice simple proof of the theorem?
Then please write it out carefully, check each detail of
the argument and each intermediate step forty-two times,
and then send the manuscript for review and publication to
the offices of the *Annals of Mathematics* in Princeton—and
(please!) not to me.

Concerning Errors

Mathematics is a human endeavor, and humans make mistakes. In the long chain of arguments of a mathematical proof,
a minor carelessness or a bit of inattention can easily become
a weak link that breaks the entire chain, and the whole thing
is worthless. A poorly rhymed couplet or a badly phrased met-
aphor may be a blemish, but it will not ruin a lengthy poem,
but a single error, no matter how small, can render an entire
proof invalid. Mathematics is without mercy and incorruptible:
almost correct is simply *wrong*.

Nonetheless, the core of mathematics is not the accumula-
tion of correct arguments but the development of ideas and
intuitions. There are, in fact, important mathematicians whose
great strength was not the creation of formal proofs but in
the development of ideas. For example, Solomon L. Lefschetz
(1884–1972), a Russian Jew who lost both hands in an industrial accident, who discriminated against Jewish students,
and who was for many years the head of the Mathematics
Department at Princeton University, was a tyrannical leader
who brought the department to a leading position in mathematics. It was said, rather maliciously, that in his entire career
he had never stated an incorrect theorem nor written a correct
and complete proof. There is a great deal one can accomplish
in mathematics with intuition—but in the end, all the pilings
have to be pounded into the ground, the knots have to be
tight, and the ship has to be secured at the pier—and for that,
there is no relying on intuition.

On the other hand, one has to admit that Lefschetz did not have it easy. What we call precision in mathematics was in his time still developing. In number theory and geometry, Euclid showed how to set up a proof properly. But when Newton and Leibniz developed calculus in the seventeenth century, they freely used "infinitesimal" numbers for computations and arguments. Euler wrote down any number of infinite products and sums without any guarantee that these represented some reasonable value. It was not until the nineteenth century that Weierstrass put the voodoo procedures of Newton and Leibniz on a proper mathematical foundation. The fields of mathematics between algebraic topology and algebraic geometry, to which Lefschetz made important contributions in the first part of the twentieth century, only got their firm foundations in the 1950s and 1960s—and for this we have to thank, among others, Alexander Grothendieck, of whom I will have more to say later.

How frequent are mistakes in mathematics? Mathematicians are human, after all, and humans make mistakes. Constantly. When we try to prove something, the process is creative and contains many errors. When I think I've proven something, I explain it to my colleagues or graduate students at a blackboard, and then I write it all down carefully, taking their comments into account. I think it through again. If no fatal flaws have shown up by that point, the thing gets sent off to a professional journal, which in turn sends it out for review by experts in the field. One hopes that any possible errors will be sniffed out during the review. What passes through these filters and finally gets published is usually pretty stable and free from errors. It is a rare occasion indeed that a major result published in one of the respected mathematical journals is found to contain an error. Of course it happens, but it is truly rare. Part of the reason is, of course, that every interesting mathematical result is not at all isolated but can immediately be checked to see if it fits other known examples, results, and theorems. If something does not fit or seems to produce contradictions to other known results, the little red alarm lights start flashing, and then mathematicians can be very persistent is searching for errors that might be hidden (however unintentionally) in the new argument.

Experience is an essential element of this procedure, for experience shows us the traps and sources of error of our business. Some foolhardiness helps also. And, of course, a blackboard. However, it is easy to convince a small group of colleagues at a blackboard of the solidity of a chain of reasoning without any of them (including the one constructing the chain) noticing the gaps or errors it contains. It was not without reason that the late Victor Klee (1925–2007), a professor of geometry at the University of Washington–Seattle, insisted, "Proofs should be communicated only by consenting adults in private." Sketches are also dangerous because one cannot tell from looking at a drawing whether it really covers all possible cases. For that reason, we mathematicians are always happy when we can translate an argument completely into text and formulas. "Driving the harpoon of algebra into the whale of geometry" is how Solomon Lefschetz put it—and he was someone who sorely needed that.

But with all that searching for errors and all the excitement that comes when something goes wrong (more of that later), we can still say that mathematics is in a state of which we can be proud. There are, after all, some very high towers of very powerful theorems resting firmly and error-free on very solid foundations.

About Computer Proofs

It is impossible to conceive of doing mathematical research any more without computers. Computers are an amazing resource and open up paths and possibilities that would not exist without them. But they are also changing the way one proves things, and in that they can be a potential and danger-ous source for errors.

Proofs come out of intuitions and ideas, and only humans have those. (At least, that is my claim. Some colleagues will disagree with me on that.) However, if assistance is offered, or at least is waiting silently to be used, one need not decline it. It is sad but true that Euler and Gauss did calculations by hand that no one today can conceive of doing without a com-puter (if they can do them at all). On the other hand, today's technology allows us to, say, generate a million examples and quickly compute the results to see if they fit, which is some-thing that Euler and Gauss could not do. Or we can generate images that can help us visualize incredibly complex situations in ways that are simply not possible without a high-speed computer and a high-resolution color monitor. In these cases, when the computer results are useful in finding the path to a proper proof, we can say a friendly "thank you" to the com-puter, write the proof down, and send it off to a journal.

But what do we do if the computer itself provides essential steps of the proof—when the necessary computations are so extensive or so complex that they could not in principle be performed by a human? I'd say we have a problem.

Can we rely on computers when we want to prove something *mathematically*, that is, with absolute 100% certainty? This is a serious and currently active question, not least because a few of the greatest successes of mathematics in the last few decades rest on feet, not of clay, but of computation. The first prominent example of such a proof was for the Four-Color Theorem, which says that any map can be colored with only four colors in such a way that no countries having a common boundary have the same color. The theorem was proved in 1977 by Kenneth Appel and Wolfgang Haken at the University of Illinois. They were able to show that any possible smallest counterexample for the theo-rem (that is, a map that *requires* five colors) must contain one of 1936 possible configurations; they then performed laborious computer calculations to exclude each of the 1936 possibilities. Such a brute-force examination of 1936 complex examples is not pretty—an anonymous critic said, "A good proof should read like a poem, this one reads like a telephone book!" But ignoring the style, can we actually rely on such proofs?

Computer-based proofs, after all, have quite a few conceivable sources of error: errors in logic, rounding errors, errors in programming, hardware errors, errors in the compilers, errors in software packages, and so forth. So maybe we can't trust such proofs. But maybe we can. One possible answer comes from attempts at providing variant proofs. In the case of the Four-Color Theorem, Neil Robertson, Daniel Sanders, Paul Seymour, and Robin Thomas submitted a clean proof whose core was a much smaller number of 633 cases that needed to be examined; listing the cases and working them through was again done by a computer program.

We are still waiting for an *elementary* proof of the Four-Color Theorem that does not depend upon computers and can be checked over by a single mathematician. Maybe such a proof does not exist. And maybe we don't need it. The other side is making progress. In 2004, Georges Gonthier submitted a proof of the Four-Color Theorem that was not intended to be understood by humans; it could only be read and reviewed by another computer program (named Coq). What he called a *formal* proof.

Is that the future of proofs: computers reviewing proofs by computers? I don't believe that, even if there are indications that suggest otherwise. After all, in 2004 we also saw formalization of an elementary proof (by Erdős and Selberg) of the Great Prime Number Theorem, followed in 2006 by the classic analytic proof. The work continues.

What's still missing is that computers start to produce ideas for mathematical propositions and then proceed to prove them. Someone who believes that this is coming and is working hard to make it so is Doron Zeilberger at Rutgers University in New Jersey. Zeilberger is a disputatious contemporary; "Dr Z's

Opinions" on his web page are always worth reading and can be inspiring. Zeilberger has become famous in mathematics for his automated methods for proving identities in combinatorics. A large number of his recent articles are coauthored by Shalosh B. Ekhad, a computer. (3B1 is a type of computer produced by AT&T; "shalosh" is Hebrew, the feminine form of the number three; "e'had" is the masculine one.) Recently Shalosh B. Ekhad has even produced articles without a coauthor.

Concerning Precision

Mathematics is not computation. Yes, this is true, and I will insist on it. There are really excellent mathematicians who never compute anything. Gerd Falting, so far the only German Fields medalist, said in an interview that he never computes examples since every time he tried to do that he was misled. There are other mathematicians who have made computation their main subject of interest—either the theory of computations (the field is known as numerical analysis or scientific computing) or actual computations, with a goal, for example, of computing the value of π to more than a quintillion places.

Does one need that kind of computation? Of what use is the billionth decimal digit of π? One might suppose that in everyday affairs a couple of digits behind the decimal point provide sufficient precision. But that's not the case. Which digits appear before and after a decimal point depends on the units one uses: 1.854 meters is 1854 millimeters; the former with three digits behind the decimal point, the latter with none. In both cases, however, we have four significant figures, and maybe we should be satisfied with that. The official distance for an Olympic Marathon is 42.195 km (or 42,195 m), which has five significant figures (at the first modern Olympic Games, the marathon was just 40 km). A list price of $1249.99 has six significant digits (and if you take care of the penny, the dollars will care of themselves). On the other hand, a colleague once remarked how silly it seems to buy a house with a nice garden for half a million and then stand in the supermarket and agonize over 69 cents or 79 cents for a container of yogurt. Maybe we really do need just three significant digits for most things. Ten significant digits would be, say, ten cents in a billion dollars. A bank accountant would not spend a long time on such a discrepancy (on the other hand, "in for a dime, in for ... ?"). But there are cases when precision is of paramount importance—for example, if you want to steer a satellite. An orbit that goes close to Jupiter to take advantage of its gravity to swing it by Saturn needs to start off in the right direction from Earth; small errors nearby lead to large errors later on, so a few cm in 100,000 km (that's ten significant digits) can make an enormous difference in the amount of fuel needed for midcourse corrections. The engineers planning these orbits need to take good care of their significant figures.

Does one also need such precise computations for proofs? Sometimes, yes. A classic example is Euler's spectacular solution of the Basel Problem, which we've already discussed. The problem concerns the sum of the reciprocals of the squares:

$$1+\frac{1}{4}+\frac{1}{9}+\frac{1}{16}+\frac{1}{25}+\frac{1}{36}+\frac{1}{49}+\frac{1}{64}+\frac{1}{81}+\frac{1}{100}+\frac{1}{121}+\frac{1}{144}+\frac{1}{169}+\frac{1}{196}+\frac{1}{225} \text{-}$$

In his essay presenting the solution, Euler provided not just one but three proofs showing that the sum is $\pi^2/6$. But the proofs could only confirm the correctness of the result. How was he able to find it in the first place? The answer is, he calculated. By an arduous computation of the value of an integral, he first computed the answer 1.644934 to six decimal places (he later extended the computation to seventeen places), an amazing feat of computation—and all with just pencil and paper, no pocket calculator, let alone a computer. But it was only at the second step that he recognized that the result is $\pi^2/6$; exactly how he did this is not clear, but Euler was an ingenious calculator. Then came the third step, proving that the infinite sum is indeed what it is. That third step took Euler several years. (He also kept his knowledge of the result a secret for that time, thereby keeping a step ahead of the others searching for a solution.)

Euler's first step is no problem for us today; with a computer it is no problem to calculate the sum to a very large number of places:

1.644 934 066 848 226 436 472 415 166 646 025 189 218 949 901 206 798 437 735

But if you do this, you also have to use considerable finesse: you have to add more than ten million terms to get the correct value for the sixth decimal place because the partial sums approach the exact result extremely slowly. For Euler's second step, *recognizing* the answer, there are today some powerful tools, but these are quite specialized. The question of whether 1.644934 can be expressed as a power of π (as well as other numbers) is, for example, nowadays answered effortlessly by an elegant program on my computer that looks for combinations of π and small whole numbers. This so-called PSLQ algorithm was discovered in 1993 by the mathematician and sculptor Helaman Ferguson.

Euler's third step, *proving* that the infinite sum is exactly equal to $\pi^2/6$ and not something just a little short of that is not easy, even today. One needs a clever idea or a considerable amount of theory. Well, OK, mathematics students in their first course in "mathematical analysis" need several hints to find a proof; students in the advanced course on analytic number theory will have learned how to prove the result quite elegantly. They will have also learned simple expressions for the sums of the reciprocals of fourth powers and sixth powers of the integers. (Euler had found these also—

"Euler, the master of us all," as his colleague Pierre Simon de Laplace wrote admiringly.)

What about the sum of the reciprocal cubes?

$$1 + \frac{1}{8} + \frac{1}{27} + \frac{1}{125} + \frac{1}{216} + \frac{1}{343} + \frac{1}{512} + \frac{1}{729} + \frac{1}{1,000} + \frac{1}{1,331} + \frac{1}{1,728} + \frac{1}{2,197} + \frac{1}{2,74}$$

Computing it to a large number of decimal places is no problem:

1.202 056 903 159 594 285 399 738 161 511 449 990 764 986 292 340 498 881 792 271 5

although one doesn't want to use just pencil and paper to do the computation nowadays. But does this set of digits mean anything? Can it be expressed in some simple fashion in terms of π or some other known constants or functions? PSLQ provides no answers—at least, not yet.

Concerning Surprises

Are there surprises in mathematics? Are there surprising proofs? One can pose that question philosophically, as the Göttingen mathematician-philosopher Felix Mühlhölzer did. Starting from Ludwig Wittgenstein's *Remarks on the Foundations of Mathematics* (1953), Mühlhölzer works out a distinction between "R-surprises" (R stands for representation), where the surprise depends only on how things are presented and "F-surprises" (F for fact), where the surprise is intrinsic to the state of affairs itself and remains no matter how well this is understood. Mühlhölzer then concludes that unlike the natural sciences, mathematics has no F-surprises; that, in the end, all of mathematics will appear simple and obvious, trivial, or even banal.

Is that really the case? That the discovery that Euler's sum equals $\pi^2/6$ will at some point lose its charm and ability to astonish? Perhaps the error lies in a misunderstanding of the words "in the end": the final and complete understanding of all of mathematics will never happen, only parts of the subject could ever be said to be complete. As long as we cannot see the whole of mathematics and can grasp all of it—and that will never happen, I maintain—there will always be surprising connections between theories and theorems that had never before seemed to have anything to do with each other.

R-surprises also arise because we learn mathematical theories and fields separately from each other (because of their history, and also because it is a practical way to structure instruction), so we do not easily see connections between them. That is why in some proofs the use of tools from a different area of mathematics, their recognized necessity and their power, can be so surprising. I'll name just one example: the question of whether one can cut a square into an odd number of triangles that all have the same area. That seems like such an elementary question that surely even the ancient Greek mathematicians must have considered it. But if a question is simple, does that make the answer easy, and is there a simple proof for the answer? Well, not yet. So far the answer seems to be no. One can cut a square into an odd number of triangles that have *approximately* the same area, but not exactly. The only proof of this, to date (by Paul Monsky in 1970) is based on a deep result concerning so-called motions in abstract algebra, and it is anything but simple.

When proofs are beautiful—and they can be, as the glistening eyes of mathematicians show again and again—there is always a little bit of surprise in play. The British number theorist G. H. Hardy in *A Mathematician's Apology* indicates the characteristics of mathematical beauty:

> What "purely aesthetic" qualities can we distinguish in such theorems as Euclid's or Pythagoras's? I will not risk more than a few disjointed remarks.
> In both theorems (and in the theorems, of course, I include the proofs) there is a very high degree of unexpectedness, combined with inevitability and economy. The arguments take so odd and surprising a form; the weapons used seem so childishly simple when compared with the far-reaching results; but there is no escape from the conclusions.

Unexpectedness, surprise: yes, they are facts in mathematics, and they provide the magic. To discover connections, results, and proofs oneself, or even to learn them from others, is a source of pleasure that has been a powerful font for millennia. It will not run dry. Gian-Carlo Rota called it "the light-bulb effect"—every so often somewhere a light bulb flashes in somebody's head. Always surprising. Sometimes very slowly, as we've learned from some of the newfangled energy-saver bulbs.

Chapter 8

Three Legends

What kinds of people are these mathematicians? These people who do mathematics not just to check the supermarket checkout slips on occasion or to review carefully the ads from the new bank credit card, and not just to look with a mathematical perspective at things in general, but who have chosen as their life's work to do mathematical research, to solve mathematical problems, to wrack their brains at mathematical conundrums. There are many stories that one can tell about these people (and their achievements), many of them true, as well as charming anecdotes and thrilling legends— and often all of these mixed together as in all historiography.

One of the classics among books about mathematicians is Eric T. Bell's *Men of Mathematics*, which was first published in 1937. It concludes with Georg Cantor, who died in 1918. So the whole thing is all rather old. Nonetheless, Bell's book is in many ways interesting and pathbreaking. One thing one notices immediately that it is indeed almost entirely about men. Only one of the twenty-eight chapters concerns a woman:

Chapter 22. Master and Pupil
Weierstrass (1815–1897), Sonja Kowalewski (1850–1891)

in which Sofia Kovalevskaya is only a "pupil" and is referred to by her nickname instead of her proper first name and by the masculine form of her last name, which she acquired as the result of a marriage of convenience (to Vladimir O. Kovalevsky, a renowned paleontologist).

As you can see, mathematical historiography may not be all that simple. But rather than trying to navigate those deep waters, I will in this chapter present three stories about mathematicians, stories I have chosen because I find them fascinating. I will try as much as possible to stick to the facts as best as I could determine them—and these are incredible enough that one hardly has to embellish them or invent alternatives.

Mathematician vs. Mathematician

In 1894 there was a remarkable public competition between two mathematicians. This was not a race to solve some mathematical problem. Such races do indeed occur, but the existence of the race is at best (or at worst) only discovered after the fact—such as the bitter and highly unfair dispute between Newton and Leibnitz as to who had the priority in discovering differential calculus. The 1894 competition was not primarily about mathematics but rather about chess. Chess is a game of pure logic; thinking counts, as do strategy, planning, and evaluation. It is thus a fertile field for mathematicians.

May I introduce the adversaries?

On the left, Wilhelm Steinitz, born in 1836 in Prague. He went to Vienna in 1858 to study mathematics. He financed his studies by getting a position as the reporter from the Austrian Parliament for the Österreichische Constitutionelle Zeitung, but soon learned that he could earn a lot more money by playing chess in coffee houses. As a result, Steinitz played a lot of chess (and, we can assume, correspondingly neglected his mathematical studies); he entered his first international chess tournament in 1862 in London. Today he is credited for revolutionizing chess theory. I don't know whether he ever completed his mathematical studies, but the game of chess owes him its "scientific approach," the systematic search for rules and patterns, and trying to assign weights and scores to fields and positions. These methods were what led him to success. He practiced *"theoria cum praxi"* (to cite the motto that Gottfried Wilhelm Leibnitz coined for the Prussian Academy of Sciences when he founded it in 1700) and won tournament after tournament. In 1886 he defeated, again in London, the Prussian Aldolf Anderssen, in a grim competition—the final score was 8:6; Anderssen (who had also studied mathematics) was an exponent of the "romantic" attacking style and was until then unofficially known as the world champion of chess. After his victory, Steinitz acquired the reputation as the world's best chess player. He dominated the game until 1894—some twenty-eight years—and actually had the first title as World Champion of chess, after defeating Johann Hermann Zuckertort, a Pole, in the first official world championship match, held in New York, St. Louis, and New Orleans in 1886.

On the right, his rival Emanuel Lasker, a German Jew, born in 1868 in Berlinchen in the district of Neumark (now Barlinek in Poland), and a brother-in-law of the poet Else Lasker-Schüler. He began his studies of mathematics in Berlin in 1889 but transferred a year later to Göttingen. In that same year, he began his career in chess with a victory at a second-division tournament in Breslau. Soon after, the fascination with chess seems to have overcome his interest in mathematics, for he interrupted his studies in 1891 and moved to London and then in 1893 to the United States.

One year later he challenged Steinitz to a match for the title of World Champion. So it came to the duel, "mathematician vs. mathematician," the twenty-five-year-old Lasker against the fifty-eight-year-old Steinitz. You are allowed to declare your sympathies at this point: for the older designated master or for the younger challenger. The underwriters of the match raised a purse of $3000, of which the victor would receive $2250 and the loser the rest. The match is "eagerly awaited in all five parts of the World," according to press reports. The *New York Times* indicates that it will report in detail about the individual games. The match starts on the 15th of March 1894, with the first set played in New York; subsequent sets are in Philadelphia and Montreal; the winner will be the first to win ten games. Lasker wins the first game, Steinitz the second, Lasker the third, Steinitz the fourth. Then there are two draws; the score is 2:2, since draws are not counted. It is a dramatic competition, the lead going back and forth repeatedly. Lasker finally wins game 15, putting the score at 9:4. He needs only one more win, but Steinitz wins game 17, in the "grandiose style of his heyday"; that game is regarded as the best of the entire match. Can Steinitz still turn the match around? Does the old man still have his bite? Lasker can't win the next game, despite some superiority; it ends in a draw.

The styles of the two competitors are actually quite similar. They both play the modern position-based strategy pioneered by Steinitz. But Lasker possibly also had a bit of psychology up his sleeve: one of the players he had massively defeated said that he did not always play the scientifically optimal move but rather the one that was most inconvenient for his opponent.

Finally, on the 26th of May 1894, Lasker ekes out a difficult win in the nineteenth game, and thus wins the match with the decisive score of 10:5 (four games had been draws).

The day after the match, the *New York Times* writes: "Lasker, therefore, is champion of the world, and deserves to be congratulated on his success, inasmuch as he has beaten his man fairly and decisively, and thereby justified the confidence which was placed in him by his backers." Lasker, whom the paper also describes as "the Teuton," was the first, and remains the only, German World Champion of chess.

The revenge match took place two years later, in the winter of 1896–1897. Lasker won decisively 10:2 with five draws; he remained World Champion for more than twenty-seven years, longer than anyone since.

As a mathematician, I am claiming Lasker as "one of us." He was not a chess player who had cut short a study of mathematics. No: not only had he wanted to be a mathematician but he was one. Indeed, after the second match with Steinitz, he stepped back from chess and continued his studies in mathematics, first in Heidelberg and then in Berlin. He did his doctoral studies in Erlangen with Max Noether, the father of Emmy Noether (1882–1935), doubtless one of the most important mathematicians of the twentieth century. His dissertation, "On Series at the Boundary of the Domain of Convergence," was only twenty-six pages long and was published in 1901. Four years later he published a long and important paper on algebra in the *Mathematische Annalen*, work that was further developed by Emmy Noether. It is clear that Lasker had ambitions for an academic career in mathematics. However, despite all efforts, he was unable to secure a suitable position in Germany, England, or the United States and thus had to continue to play chess. Perhaps Lasker is the real predecessor to the fictional pianist who appears in Wolfgang Hildesheimer's *Lieblose Legenden* (from 1952) and who *actually* wanted to be an insurance salesman but whose stern father forbade him his dream occupation and forced him into an undesired but brilliant career.

Albert Einstein got to know Lasker fairly well in Princeton, where they went for walks together. In 1952—the year that Hildesheimer's *Loveless Legends* was published—Einstein provided a preface for the English translation of a biography of Lasker and included the following:

> To my mind, there was a tragic note in his personality, despite his fundamentally affirmative attitude

towards life. The enormous psychological tension, without which nobody can be a chess master, was so deeply interwoven with chess that he could never entirely rid himself of the spirit of the game, even when he was occupied with philosophic and human problems. At the same time, it seemed to me that chess was more a profession for him than the real goal of his life. His real yearning seems to be directed towards scientific understanding and the beauty inherent only in logical creation, a beauty so enchanting that nobody who has once caught a glimpse of it can ever escape it.

Spinoza's material existence and independence were based on the grinding of lenses; chess had an analogous role in Lasker's life.

Lasker had, in fact, entered Einstein's domain by publishing an article critical of the special theory of relativity, questioning whether the speed of light in a vacuum is indeed finite and universal.

Einstein concludes:

But I liked Lasker's immovable independence, a rare human attribute, in which respect almost all, including intelligent people, are mediocrities. And so I let matters stand that way.

Was It Kovalevskaya's Fault?

Why is there no Nobel Prize for mathematics? There are several legends arrayed around this question—some of which are considerably more interesting and exciting than the rather prosaic truth.

For example, from an article written by Jeanne Rubner in 1999 in the *Süddeutsche Zeitung*:

> The recipients of the Fields Medal derive some of their limited fame from the recognition that the honor is treated as a covert "Nobel Prize" for mathematicians. The non-existence of that prize, by the way, is due to a talented mathematician. The Russian Sofia Kovalevskaya, who would have certainly warranted a Nobel Prize for Mathematics, did not want to become Alfred Nobel's lover. He got his revenge by not including mathematics in his will some 100 years ago.

So, it's Kovalevskaya's fault.

But there are also other versions of the story. The French version says that Gösta Mittag-Leffler, one of the most eminent Scandinavian mathematicians, had an affair with Nobel's wife, identified as Ms. Kovalevskaya, and that Nobel therefore wanted to prevent his getting the prize. This is, I must admit, my favorite version of the story—for once the mathematician is not the loser nerd but the Romeo. It is, however, not really plausible, because Nobel actually never married: he had motherly friends and patronesses and correspondents, but no wife and probably no mistress.

Was Nobel interested in Kovalevskaya? The German *Wikipedia* reports:

> At the end of 1887 Sofia met Alfred Nobel, who also courted her, but an affair between the two never developed. Until today the rumor persists that there is no Nobel Prize for Mathematics because Kovalevskaya had an affair with Nobel and then left him for Gösta Mittag-Leffler. There is no basis for this

rumor because Sofia Kovalevskaya did not have an affair with Mittag-Leffler either.

One might take issue with the somewhat snarky tone of these comments, and one might also ask how the Wikipedia columnist came to learn all these salacious details.

Yet another version leaves the women out of the picture and just speculates that Nobel and Mittag-Leffler did not like each other for some reason, and Nobel did not include mathematics among the prizes he endowed because he thought Mittag-Leffler, in view of his prominent position in the Swedish Academy, was a likely recipient of the prize, and he wanted to prevent this.

Indeed, Gösta Mittag-Leffler (1846–1927) was a prominent member not only of the Swedish Academy but also of Stockholm society. He was a renowned scientist, corresponding with mathematicians all over Europe; he was wealthy and had a position in society; he built himself a marvelous villa in Djursholm, a Stockholm suburb, for which he was even able to arrange the extension of a streetcar line. In society, he was generally surrounded by three women: his wife, to whom he owed a substantial part of his wealth; his sister Anne Charlotte Leffler, an author who had a great success with, among others, her stories "From Life" about the upper circles of Swedish society; and Sofia Kovalevskaya, called Sonya, who was (not without cause) the subject of much fantasy.

Sonya Kovalevskaya was born (as Sofia Korvin-Krukovskaya) in 1850 in Moscow; her father was a Russian general, her mother came from a German aristocratic family. After her father retired from the army, the family moved to a large estate in what is now Belarus. The walls of her room helped bring young Sonya to mathematics: in the absence of other wallpaper, her room was covered with lecture notes for differential calculus from her father's student days. For the eleven-year-old Sonya, this wallpaper opened a window into a strange, fascinating, and, for a long time, totally incomprehensible world; it was also a welcome diversion from the social requirements of her parents' salon, from fights with her seven-years-older sister, and many other dislocations. We know all this from Sonya herself, because she wrote it down. Her memoir, *Vospominaniya detstva*, was published in 1893, translated into Swedish, German, and English (as *A Russian Childhood*), and became a best-seller; it was forgotten, but rediscovered, and is a literary gem. It gives Sonya the first word about the legends of her upbringing; the second place belongs to her friend Anne Charlotte Leffler, who wrote her biography.

Sonya, if one trusts her own description, was a stubborn child, rejected by her mother. The novelist Fyodor Dostoyevsky was a frequent visitor to the house, having fallen for Sonya's older sister, Anjuta. Sonya developed a serious crush on him: "I completely fell under his influence." The German novelist Peter Härtling describes the situation succinctly: "Teen girl rivalry ends in sob-filled nights." She seems to have gotten over it somehow, because at age eighteen she married Vladimir Kovalevsky, probably a marriage of convenience so she could leave Russia to pursue her studies in the West. And this she did successfully. She studied mathematics at Vienna and Heidelberg, but then transferred to Berlin to study with Karl Weierstrass, the "father of exact analysis." Because at that time a woman could not be admitted as a student in Berlin, Weierstrass gave her private lessons for four years. Her thesis was submitted to the University of Göttingen (more liberal than Berlin), and her PhD was granted, without the usual oral examination, with distinction, in 1874; she was twenty-four.

The next years were turbulent, with repeated trips between Berlin and Moscow in vain attempts to be recognized and obtain a position in her native country, as well as unsuccessful attempts to be a mother and homemaker. By 1880 she returned to mathematics and lectured on the results in her dissertation at a large congress in Moscow. In 1882 she was invited by Mittag-Leffler (who had also been a student of Weierstrass) to visit Stockholm; a year later she obtained a position at the University of Stockholm, and in 1884 she became a regular professor at the university (after considerable lobbying by Mittag-Leffler, against strong opposition from the faculty). Kovalevskaya was thus the first female mathematics professor at any European university. In 1888 she received the Prix Bordin of the Paris Academy of Sciences; in 1899 she was granted tenure at the University of Stockholm. Unfortunately she died two years later from pneumonia, at the age of only forty-one.

So why is there no Nobel Prize for mathematics? Today we know that Alfred Nobel and Gösta Mittag-Leffler knew each other and probably did not like each other very much. If Sonya Kovalevskaya had anything to do with that, it is not clear. But we do know of some testy exchanges between Nobel and Mittag-Leffler during the struggle for a full professorship for Kovalevskaya. Mittag-Leffler wrote to Nobel asking for financial support for the professorship, since otherwise she might have to return to Russia after the expiration of her temporary appointment. Nobel answered dismissively that it would be just as well for her, since mathematics in Stockholm was after all rather provincial.

But does this have anything to do with the "missing" Nobel Prize? Probably not. Nobel is likely to have not considered mathematics as a science that provides the "greatest benefit on mankind." In his will, this is the decisive criterion for the awards. So there are prizes only for physics, chemistry, medicine, literature, and peace. The economics prize is a very late addition to the group, endowed by the Swedish Riksbank to celebrate its 300th anniversary in 1968; some circles in the Swedish Academy of Sciences consider it a mistake. In any case, it's not a "real" Nobel Prize.

Once the Nobel prizes were established, Mittag-Leffler involved himself thoroughly. For example, he made sure that the Nobel Prize for Physics in 1903 for work in radioactivity was awarded not only to Henri Becquerel and Pierre Curie but also to Marie Curie. He wrote Pierre Curie and asked him for confirmation that his discoveries were made in joint researches with his wife and that they should thus be jointly recognized. Mittag-Leffler submitted the resulting letter to the academy and thus ensured that Marie Curie was included in the award. A mathematician as a promoter of women? Perhaps. But in this case perhaps simply a promoter of excellent science. Mittag-Leffler also tried for many years to get an award for Henri Poincaré, starting in 1901 (the first year the prizes were awarded) and continuing until Poincaré's death in 1912. After this, Mittag-Leffler's interest in the Nobel Prizes waned considerably—perhaps in part because he was not able to convince the academy that the physics prize might occasionally go to a mathematician who had made great contributions to the field.

But we can also answer the question, "Why is there no Nobel Prize for mathematics?" in an entirely different way: we don't need it. We have the Fields Medal, which was first awarded in 1936, and since 1950 has been awarded every four years at the International Congress of Mathematicians. Admittedly, they are not as lucrative as the Nobel Prize, providing only 15,000 Canadian dollars. And, unlike the Nobels, they can only be awarded to active mathematicians under the age of forty (no "prizes for geezers").

Another answer is: mathematicians don't want it. In the 1960s, prominent mathematicians in the Academy of Sciences in Stockholm were asked if there was any interest in a Nobel Prize for mathematics. At the time, the answer was a sort of clear "no": leading Scandinavian mathematicians, including Lars Gårding, Lennart Carleson, and Lars Hörmander, thought that a competition or race for a Nobel Prize would be detrimental to the cooperation among mathematicians that was so important to the progress in the field. They therefore declined, with thanks.

And third answer: there is actually a Nobel Prize for Mathematics. Since 2003 the Abel Prize has been awarded annually by the Norwegian Academy of Sciences, which also awards the Nobel Peace Prize. The monetary award is handsome (6 million Norwegian kroner, a little over a million dollars) and is given for "outstanding scientific work in the field of mathematics," which has meant a lifetime achievement (so it's for older mathematicians). The king of Norway hands out the prize in person—altogether, a prestigious prize. It's not called a "Nobel Prize," but then, we don't know if Alfred Nobel would have thought that would be a good thing.

The Disappearance of Alexander Grothendieck

Alexander Grothendieck is one of the most important, intimidating, and mysterious mathematicians of the twentieth century. His origins, his biography, his achievements, and his "disappearance" are all surrounded by legends. His father was a Russian anarchist of Jewish descent named Alexander (Sasha) Shapiro; during the chaos of the Russian Revolution, Sasha was twice condemned to death and twice escaped. He fled to Hamburg and worked there as a street photographer. In Hamburg he met the German journalist and author Johanna Grothendieck, whose autobiographical novel *A Woman* is judged by those who know it to be the equal in style and substance to Alfred Döblin's *Berlin Alexanderplatz*; it was unfortunately never published. Alexander was born in 1928. After his parents fled Germany to escape the Nazis, he was in part raised as a foster child by a priest and his family. He later followed his parents to France. In 1940 the entire family was detained in a concentration camp by the Vichy regime. His father died in 1942 in Auschwitz. Mother and son were separated but survived and were able to find each other after the camps were liberated by the Allies. Despite all the turmoil of war, camps, and flight, Alexander was able to complete his baccalaureate in 1945 and went on to study mathematics at

Montpellier—one of the most hidebound, lifeless, and calcified mathematics institutes in France. There he worked with great energy and effort on a comprehensive theory of integration, which, however, was already long known elsewhere. In his memoir *Récoltes et Semailles* (*Reaping and Sowing*), Grothendieck later wrote about his time in Montpellier:

> When I think back upon these three years, it is clear to me that I did not in the least waste my time there. Without being aware of it, I learned to appreciate the isolation that is so essential for work in mathematics. Without having heard it from anyone, and without ever meeting anyone who might have shared my thirst for knowledge, I could sense that I would be a mathematician, someone who "makes" mathematics, as one "makes" love.

In the fall of 1948, at the age of twenty, Grothendieck came to Paris, and thus into the center of what was happening in mathematics. It must have been a revelation to him. Mathematics in Paris was fermenting, lively, and productive. Grothendieck learned from world-famous mathematicians such as André Weil, who then recommended him, in 1949, to Laurent Schwartz in Nancy. Schwartz was at that time already a mathematical superstar, dominating the important field of functional analysis, and recognized with a Fields Medal at the International Congress of Mathematicians in 1950 for his *Theory of Distributions*. When Grothendieck came, from Montpellier via Paris, to study with him, Schwartz gave him a technical essay to read, an essay that concluded with fourteen problems that Schwartz and his coauthor Jean Dieudonné had not been able to solve. Amazingly, Grothendieck solved half of these within a few weeks and all of them by the end of the year. All of them! Schwartz, who became his thesis advisor, described his work style thus: "He lives alone and works 25 or 26 hours a day."

Grothendieck completed a 300-page dissertation based on these solutions in 1953; Schwartz called it a "masterpiece of immeasurable value."

After these fourteen drum rolls in functional analysis, Grothendieck turned, in the fall of 1954, to other fields of mathematics: complex analysis, algebraic geometry, and homological algebra. His achievements caused yet another sensation. In 1957, at the first of the now-legendary Bonner Mathematische Arbeitstagungen (Mathematical Workshops in Bonn, begun by Friedrich Hirzebruch), Grothendieck presented his version of the Riemann–Roch Theorem; he lectured for two hours on each of four days to a spellbound audience. He was not yet thirty years old and already counted as a giant of mathematics. It was said that other famous mathematicians, such as Michael Atiyah and Stephen Smale (no mathematical slouches either, with Fields Medals and impressive results), had paled to "perfectly ordinary geniuses" next to Grothendieck.

After spending two years in Brazil and Kansas, Grothendieck was appointed to the new Institut des Hautes Études Scientifiques (IHÉS; Institute of Advanced Scientific Studies) in Paris. The institute, founded in 1958 by a private benefactor, soon acquired an international reputation, in no small part because of Grothendieck. Here again, he achieved amazing results. He became an internationally renowned (and feared) monument of mathematics, with monumental achievements and monumental expectations. His multivolume *Élements de Géométrie Algébrique* provided an entirely new foundation for algebraic geometry that was further elaborated by his *Seminaire de Géométrie Algébrique*, documented in multiple typewritten volumes. These two sets of volumes, known as EGA and SGA, remain the foundations of one of the central fields of modern mathematics, and anyone who wishes to do research in the field has to work through the bookshelf-worth of EGA and SGA.

In 1970 came a great turning point, as Grothendieck himself termed it. In a fight resulting from the discovery that some of the funds for the IHÉS came from the French defense ministry, Grothendieck quit. He left his position at the IHÉS and step by step removed himself from mathematics; he also left his wife and family and turned aggressively against former friends and colleagues. He joined the peace, antinuclear, and ecology movements; he became a Buddhist and has written thousands of pages of *Meditations*. Since 1991 he has lived in seclusion in an unknown place in the south of France.

Winfried Scharlau, a retired professor of mathematics at the University of Münster, is working on a comprehensive biography of Grothendieck. The first volume (of a planned three) was published in 2008. Scharlau succeeds in clearing up, at least in part, many of the puzzles and legends that have surrounded Grothendieck's origins, life, and disappearance. But many questions remain. As an example: on March 28, 2008, *Die Zeit* published a full-page encomium on the occasion of Grothendieck's eightieth birthday, written by Scharlau and entitled "In Higher Dimensions." I happened to be talking with the head of a division of the German Federal Ministry of Research related to the 2008 "Year of Mathematics," and he mentioned that the article in *Die Zeit* appeared suspiciously in the issue for the week that included the first of April. And thus he asked, quite seriously and totally justified, "Does this guy really exist?"

Chapter 9

What Kinds of People Are These?

Do you know any mathematicians? They do exist in real life, with interesting biographies, but also in novels, where they are sometimes so strange as to be unrecognizable. Who can tell them apart? The Carl Friedrich Gauss from the biography by Hubert Mania is a historical genius; the Karl Friedrich Gauss from the novel *Measuring the World* by Daniel Kehlmann is an interesting but somewhat bizarre novelistic invention.

Mathematicians in novels are probably always a little removed from reality—like, for example, Ulrich from Robert Musil's *The Man without Qualities*:

> About Ulrich, on the other hand, one could say one thing with certainty, that he loved mathematics because of the people that could not stand it. He was not so much scientifically but rather humanly in love with science. He saw that in all the questions it dealt with, it thought differently from ordinary humans.

But mathematicians in real life are not novelistic inventions. They are real. Normal people. Sometimes they cry. Sometimes they laugh. Not all of them are men. Not all are geniuses (and even the geniuses aren't always recognizable as such). The important mathematicians are power tools with a passion for precision and the willingness and stamina to follow through with things. But, for all that, still ordinary people.

I will recount some stories about a few of them. They will not be heroes of history, whom I know from encyclopedias or textbooks, but rather people I know or have known, colleagues, interesting contemporaries. Even if I have to admit (through clenched teeth, and as quietly as possible) that most mathematicians are actually rather boring—just like most politicians, most lawyers, and most journalists. But telling stories about the most boring, or even just the most average, of my colleagues is not a winning concept. What I am interested in (and probably you as well) are the careers that go a little outside the ordinary paths. In this chapter I present a completely subjective and somewhat arbitrary selection. So I will not tell you about Wendelin Werner, a Fields Medal honoree, and how he came to a major role in Romy Schneider's last film—although you might find that interesting (the film, in case you need to know, is *The Passerby*, a French film from 1982 whose original title was *La passante du Sans-Souci*). Nor will I tell you about Victor Mal (warning: pseudonym), my fellow student and sometime roommate at MIT, a Belgian who went to Princeton after concluding his studies, where he married one of his students. Subsequently he held two professorships simultaneously, in San Diego and in Salt Lake City, and used the San Diego computer system for running a dating website. After his involuntary separation from the university, he built the web portal www.playboy.com and became the "Senior Vice President for New Media Technology" at Turner Broadcasting/AOL Time Warner. Today he has his own company that does direct marketing on the Internet. An interesting career, but not one I'll tell you about.

Paul Erdős: Traveler

The science journalist Bruce Schechter begins his biography of Paul Erdős with a confession: the mathematician Ron Graham had been telling Schechter about Erdős since 1982; Schechter had tried to meet Erdős again and again, but never succeeded; when he learned in 1996 that Erdős had died, he decided to write Erdős's biography. He started by talking with the many friends and colleagues from all over the world who had come to the funeral service in Budapest.

I think this story does not speak well for Schechter, but it also says a lot about Erdős—perhaps more than many of the anecdotes about him that are in circulation. Erdős was in fact not at all difficult to meet. He traveled constantly for decades. One could meet him at conferences all over the world. Or one could simply wait in some place where there were mathematicians and wait for him to come by. I can assure you that sooner or later he *would* come by. He came, he drank coffee, gave lectures, talked and worked with old and young colleagues, discussed problems, did research, played.

Erdős came from a Jewish family in Budapest. Two older sisters died of scarlet fever just before Erdős was born in 1913. He thus grew up as an only and coddled child, home-schooled by his parents, both teachers. His talents were recognized and encouraged early on—a child prodigy in a golden cage, in contact with other highly gifted children already at a young age. He was a part of an amazing generation of mostly Jewish Hungarian mathematicians and physicists who contributed to an amazing variety of fields—ranging from fascinating play in the universe of accidental and extreme graphs (Paul Erdős, Alfréd Rényi, Paul Turán) to the development of modern physics and the atom and hydrogen bombs (Eugene Wigner, 1902–1995; John von Neumann, 1903–1957; Edward Teller, 1908–2003). The political disturbances, the crimes against humanity, and the stupidities of the twentieth century also determined Erdős's life. Among them the Horthy regime in Hungary, anti-Semitism, the Second World War, the isolation of Eastern Europe under Stalin, and the McCarthy era in the United States with its persecution of communists and imagined communists.

Erdős produced at the age of twenty (still in Hungary) a brilliant, elegant, extremely simple, and new proof of the Bertrand Postulate, which says that between any number and its double there is at least one prime. This became the core of his doctoral thesis in 1934. Fifteen years later, already in the United States, he garnered recognition for a completely new proof of the Great Prime Number Theorem. This theorem had been proven around 1900 with methods from differential calculus; Erdős and the Norwegian Atle Selberg were able to prove the theorem in an elementary way without resorting to calculus. Unfortunately there was an argument between Erdős and Selberg concerning how the proof came together and how to apportion the credit. Selberg ultimately published his version of the proof himself and thus reaped most of the fame, including a Fields Medal. Erdős left Princeton and never held a formal position or fixed location after the 1950s. He traveled from country to country, attended conferences, was a guest of many friends and colleagues on all continents, and worked with dozens of them simultaneously on many different problems.

Erdős was always on the move, but always approachable. I met him on numerous occasions and have heard him give many talks—he gave wonderful talks, full of humor, made fun of himself as an ancient man, and he posed problems for which he set prizes ranging from $5 to $5000. I no longer remember when or where I first met Erdős. In any case, I encountered him memorably at a conference in Oberwolfach in November 1992. At that time, he was a little old man, kind, friendly, interested, and approachable—approachable even for me, when I was far from being an established mathematician. Later, in 1995 and 1996, we sat together in Berlin, and he made suggestions for proofs that we should write down for a book we were planning about "The Book of Proofs." He had a complete collection of the relevant sources in his head, he knew in which Hungarian journal and in which year a certain proof was published—and he knew the associated history and the relevant stories. He talked about the sixty years of mathematics history that he had participated in, obviously with great pleasure.

I believe that Erdős really succeeded in making his scientific work into play. His toy box and tool kit contained numbers, figures, configurations, and other simple objects. He searched for extreme configurations and thus became the founder of extremal graph theory. He asked about unavoidable patterns in apparently random structures and thus made important advances in Ramsey Theory. And he recognized the importance of chance, which led him, together with Alfréd Rényi, to develop Random Graph Theory, starting in 1959. That theory is based on the recognition that pure chance can lead to patterns that would be difficult or impossible to construct according to a plan or method. Especially if one is trying to direct chance.

Thus one can ask, for example, if there is in every large set of people a group of six people who all know each other or of whom none knows any of the others. A theorem from 1930 by the British mathematician and logician Frank P. Ramsey says that in *every* sufficiently large set there is such a group of six. How large, then, is "sufficiently" large? Ramsey's inductive proof shows that we can be certain that there is such a group if there are more than 252 people in the set. Erdős's probabilistic argument shows that 17 people are not enough: if the connections among 17 people are governed by chance, there is a finite probability (almost 25%, as one can compute fairly readily) that there are no groups of six who either know each other or know none of the others.

But how large, at a minimum, is the set of people for which one can be certain that there are six people who know or don't know each other? This is called the Ramsey Number R(6,6), and no one knows what it is exactly; it is known to lie somewhere between 102 and 165. With his characteristic humor, Erdős said,

> Let us suppose that Aliens conquer the Earth and threaten to destroy it within a year unless people can come up with the Ramsey Number for groups of five. We could then put the best minds and the most powerful computers together and probably come up with the value within a year. If, on the other hand, they want the value for groups of six, the only option is to start a counterattack.

Paul Erdős, 1913–1996, what kind of a life did he lead? Certainly not an ordinary one. He never had a fixed domicile; the only fixed reference points were his mother's apartment in Budapest (but not after her death in 1971) and the guest room at the house of his close friend in Princeton, Ron Graham. His possessions fit into a small, brown, scuffed leather suitcase. They consisted of a couple of silk shirts, underwear, toiletry kit, notepads, and almost nothing else. Erdős traveled light. But the fact that he was constantly traveling and still was able to write more than 1500 papers with more than 500 coauthors makes it clear once again that mathematics needs no fixed location.

Other than the colleagues with whom he worked and a few other friends, he probably had no other close relationships. He said that he had tried sex once, at the age of seventeen, but didn't particularly like it and therefore never tried it again. (Friends who knew him well said that this was probably true, but that the age was likely considerably older than seventeen.)

But Erdős was totally anchored in the "real world." I remember, on his last visit to Berlin, a Sunday dinner with a pot roast and (Austro-Hungarian) dumplings at my colleague Martin Aigner's. "Uncle Paul" conversed in German and English as the conversation flowed through politics, arts, and, of course, mathematics. During the discussions, Paul also took a couple of phone calls, dealing with one in French, another in Hungarian. Indeed, we can imagine Paul as a happy man.

Gian-Carlo Rota: Provocateur

Gian-Carlo Rota was one of my professors at MIT. An Italian, he was born in 1932 in Vigevano, south of Milan. Nino Rota, who composed the music for films by Fellini (*La Dolce Vita*, *8½*), Visconti (*The Leopard*), and Coppola (*The Godfather*), was an uncle. His father, Giovanni Rota, was a civil engineer, specializing in earthquake-proof structures, and an antifascist who was said to be on Mussolini's hit list. As a result, the entire family fled into the mountains, followed by a dramatic game of hide-and-seek with the authorities, falsified identities, and the discovery and arrest of the father. Fortunately there was a bit of confusion, as an engineer named Rosa was highly valued by the fascists, allowing the family to escape to Switzerland. In 1946 they were able to immigrate to Ecuador, where Gian-Carlo finished high school. He then went to the United States and did his undergraduate work at Princeton; he received his PhD from Yale at age twenty-four, working with Jacob T. Schwartz on functional analysis. After intermediate stops as a postdoc in New York and at Harvard, he arrived in 1959 at MIT, the legendary university in Cambridge, Massachusetts, at which radar and stroboscopes were among the many proud inventions. Rota was professor of mathematics there for almost forty years, from 1959 until his death in 1999, with only one break, 1965–1967. In addition, he was for many years a consultant at Los Alamos National Laboratory (the lab that developed the atomic and hydrogen bombs), where he worked with Stanisław Ulam, whom we have already met in one of the coffee houses. He was awarded a medal by the National Security Agency. Of course, I do not know what it was for.

Rota started out specializing in functional analysis, at the time an active, respected, and modern area of research. But in 1964, Rota produced a bombshell: with remarkable self-assurance, the thirty-two-year-old Italian published a paper with the title, "On the Foundations of Combinatorial Theory I. Theory of Möbius Functions." Nine further papers entitled "On the Foundations of Combinatorial Theory," all with coauthors, followed, starting in 1970. It needs to be said that at the time, combinatorics was not at all in the ranks of the respectable, worthy fields of mathematics. It deals with finite structures, graphs, sets, and counting problems. Even though giants such as Gauss and Euler had made contributions to the field, it was in 1964 considered a playground for students, a proving ground for problem solvers, a field with many clever tricks but no substantial theory. Rota apparently had the idea to change this impression. And he succeeded. When he received the renowned Steele Prize from the American Mathematical Society in 1988, the AMS pointed out that the field

> was now highly active and universally accepted.
> Two of its most prominent features are its unifying techniques … and its deep connections with other branches of mathematics.

The 1964 paper was specifically cited as

> the single paper most responsible for bringing on this revolution.

Rota had brought the field from the scientific dustbin to the spotlight.

Combinatorics is also known as "discrete mathematics," because it deals not with continuous and smooth structures (the subject of differential calculus, for example) but with finite numbers of individual ("discrete") objects. Of course, the concept is a little ambiguous. And, of course, people who like a little wordplay discuss topics in "indiscrete mathematics." Rota not only accepted that but also played with it himself: his two later volumes of essays were *Discrete Thoughts* (1986) and *Indiscrete Thoughts* (1997). A third volume of *Forbidden Thoughts* was planned, but he was not able to complete it.

I attended many lectures by Rota. They had a style. His best lectures were the ones for which he had prepared badly or not at all: in these lectures he did not cover a lot of material but he improvised brilliantly, with witty and charming discussions of the history of mathematics, philosophy, or a thousand other topics. I learned a lot from him—for example, that the blackboard must be immaculately clean before the lecture starts, because accidental chalk marks can easily distract the audience. Or that each lecture has to have a central message, "something to take home." Rota also claimed that it is a good idea to publish the same results or insights repeatedly, but each time newly developed and improved. Or that small errors and the occasional imprecision are not really serious. But in making that suggestion, he may have been making use of Oscar Wilde's dictum, "I always pass on good advice. It is the only thing to do with it. It is never of any use to oneself."

Gian-Carlo Rota was a provocateur in a suit. I only know him in a suit and tie. He was, after all, not only a mathematician but also Italian and a philosopher. He advocated the philosophy of Husserl and Heidegger, and thus provided a strong contrast with the MIT philosophy faculty, which was dominated by the analytical philosophy of Russell and Wittgenstein. He had to fight to have his philosophical bona fides recognized. It took an intervention by the president of MIT before Rota was able to become "professor of applied mathematics and philosophy," and thus able to teach courses in both departments that would count toward a degree. His philosophy lectures were immensely popular. For many years, they were held on Fridays at 6 p.m., apparently in part to reduce the number of students enrolled in the course.

In January 1999, I tried to get him to write an essay on gay mathematicians for the *Mitteilungen der DMV* (the *Notices of the German Mathematical Society*), for which I was the editor at the time, inviting him to be as provocative as he liked. He wrote back that he did not need to make any more enemies. Because of an essay concerning the "corrupting influence of mathematics on philosophy," the analytic philosophers had already held him to be an incarnation of the devil; in that connection, he told me about a philosophy lecture at MIT at the end of which half the audience applauded and the other half sat on their hands. I don't know of any such extremely polarized reactions to lectures in mathematics, not at MIT or elsewhere. Apparently the emotions don't run as high in our field.

That Rota was an essayist with a sharp tongue was also demonstrated in the journal *Advances in Mathematics* that he founded in the 1960s and edited dictatorially: articles were in general not reviewed externally; Rota decided what to publish based on his own tastes and inclinations. Many issues closed with book reviews by Rota, reviews that never lacked for pepper and spice. He dealt with the entire collected works of seven giants of mathematics (Emmy Noether among them) in a single page. A book about contemporary philosophers he dismissed in a single sentence: "If pygmies cast such long shadows, it must be late in the evening."

Gian-Carlo Rota was found dead in his apartment in Cambridge on April 18, 1999. He was said to have had a heart attack. However, I know that his death cannot have been a purely unexpected event because I am told that the day before his death, Rota cleaned up his office. He sorted through piles of papers, arranged books in order, and so forth. But why? He had not done that for decades. I remember his office, room 2-351 in the main building of MIT, from the time when I discussed my ideas on the way to my PhD dissertation with him. He noted the ideas in thick notebooks and discussed them with me animatedly and encouragingly. We sat in the middle of the labyrinth made by stacks of books along with piles of journals, manuscripts, and reprints—a labyrinth that had grown higher and denser over decades; clearing it up would clearly be a Sisyphean task that could never be concluded.

Persi Diaconis: Magician

Persi Diaconis was born in 1945 in New York to a family of musicians. Already at the age of five he was taking violin lessons at the Julliard School, which he continued until he was fourteen and left home. At the age of five he also got his hands on Howard Thurston's book *400 Tricks You Can Do* and was captivated. He started frequenting a magic-supply store and met similarly inclined spirits, including Martin Garner, the legendary author of the "Mathematical Games" column in the *Scientific American*, and Dai Vernon, at that time one of the most important magicians anywhere—and four times as old as young Persi. When he was fourteen, Persi left school and home to travel with Dai Vernon as his assistant, hitchhiking from town to town. When Vernon settled in Los Angeles, young Diaconis set off on his own. In Chicago he was making $50 a night performing as a magician, but ultimately he drifted back to New York.

By coincidence he met a friend from his school days at the big Barnes and Noble store in Manhattan. The friend had in the meantime become a mathematician, and enthusiastically recommended William Feller's *Probability* to Diaconis. It was the greatest book, he said. As a consequence, Persi started taking evening courses in mathematics at City College in New York in order to be able to make his way through Feller's tome. He finished with a bachelor in mathematics two and a half years later. A somewhat unconventional letter of recommendation from Martin Gardner to a statistics professor whom Gardner knew got Diaconis into grad school at Harvard University:

> Dear Fred
> I am not a mathematician, but of the ten best card tricks that have been invented in the last five years, this guy Diaconis invented two of them, and is interested in doing statistics. He really could change the world. Why don't you give him a try?

A few years later, Diaconis got a PhD with a dissertation on number theory ("The Physics of Numbers") and started his second career as a professional mathematician—but with a thoroughly unconventional field of interest and an equally unconventional working style, and some extraordinary results. One of the problems he worked on was: How often does one have to shuffle a deck of fifty-two cards to make sure the cards are in a random order? The answer is surprisingly exact: if you use a dovetail shuffle, you must go through exactly seven shuffles. The theory leading to that result is difficult and uses what is called the representation theory of finite groups. (The rearrangement or shuffling of cards is a permutation of the original ordering, and the permutations of a finite number of objects form an algebraic structure called a group. The theory of groups is an important part of algebra.) So here we have a mathematician with a rather abbreviated school career who uses complicated tools of theoretical algebra to solve problems in magic, card shuffling, statistics, and probability theory. And when he's so inclined, he will conclude a technical lecture with a demonstration of a "perfect shuffle," which returns the deck to its original order after several repetitions.

But does he really do a lot of mathematics? Well, yes, but he is not one whose work depends on immense amounts of mathematics stored and shelved in one's head. There are some mathematicians who, when they need a result or a formula, will quickly derive it from what they know, and will feel empowered and proud having done that. But there are other mathematicians who will simply look up the result in some reference, and are correspondingly proud of their search techniques and bibliographic knowledge. Persi Diaconis is the latter kind of mathematician.

A dual talent: magic and mathematics. Diaconis has used this unusual combination to expose charlatans and to uncover cheating in casinos. He has successfully debunked self-proclaimed parapsychologists. Unmasking the legendary spoon-bender Uri Geller was one of his successes.

Diaconis is currently a professor of statistics at Stanford University, as is his wife, Susan Holmes. He continues to work with games of chance, coming up, for example, with surprising results about fair dice or about the predictability of coin tosses. He continues to explore new worlds. To my surprise (and probably also his), we were cochairs in 2006 at a semester program on "Computational Applications of Algebraic Topology" at the Mathematical Sciences Research Institute (MSRI) in Berkeley. Topology is a huge and difficult area of mathematics of which I really know only some small parts (and Diaconis even smaller parts); until recently it had the reputation of having no applications at all. Nonetheless it turned out to be a lively program—helped, of course, by the location, a remarkable institute overlooking San Francisco Bay from high up in the Berkeley Hills. In conversation, Persi Diaconis is witty, lively, and imaginative. I'm looking forward to learning about his next expeditions into the areas between mathematics, chance, magic, and reality.

Daniel Biss: Politician

I met Daniel Biss only once, in April 2001 in Berkeley, where he gave a seminar talk on a spectacular result from his doctoral research; he completed his dissertation a year later. The talk was interesting and accessible, treating the general area of his research and the significance of his proofs, but he did not go into the technical details. The subject is something called spaces of oriented matroids, that is, some highly abstract discrete models for geometry. We should not discuss that in too much detail here. (Several years ago, some colleagues and I wrote a book on this model that's more than 500 pages long.) The question was, how good is this model? Biss's answer: good enough to serve as a foundation for further research, which is what had been hoped for. At the time, Biss was twenty-four years old, a young math star with a small face surrounded by black curls. He comes from the Midwest of the United States, born in Akron, Ohio, and raised in Bloomington, Indiana, where his parents, both violinists, taught at the university. As a child, he also learned to play the violin but found mathematics to be even more attractive. He was encouraged by math professors at the university but also learned to love math at summer research projects organized by "Research Experiences for Undergraduates." In 1995 he went to Harvard for his undergraduate studies, and in 1999 went on to MIT for his graduate work in mathematics. At MIT he studied with Robert MacPherson, whom I admire greatly and who was one of the reviewers of my own postdoctoral thesis. In 2002, Biss was awarded a research fellowship by the Clay Foundation, which supported him for five years. (Terry Tao, the "Mozart of Mathematics" whose work on prime numbers was discussed earlier in this book, for example, also received this fellowship.)

At the age of thirty, in 2008, Biss was appointed an assistant professor at the University of Chicago, and he even held lectures there, but his main energies were focused on electoral politics. Already in 2004 he had worked for the presidential campaign of John Kerry, but in 2008 it was his own campaign for a seat in the Illinois State House of Representatives that he was working on. He ran his campaign without a large office and without much support from the Democratic Party, but with enthusiasm and very effective Internet marketing. He was able to raise $435,000 in campaign contributions, and the *Wall Street Journal* ran a report on the young candidate fearlessly competing against the three-term Republican incumbent. He almost made it. In November 2008 he got 48.5% of the vote in his district—presumably helped to some extent by Obama's victory—compared to his opponent's 51%.

In January 2009 Biss celebrated his entry into politics as "a hell of a good start" and commented as an aside that he had also as good as ended his academic career, since he had not applied for tenure at his university. And while the electoral defeat was indeed the start of a career in politics, in mathematics he had recently suffered some stinging defeats.

In fact, Biss should have had an excellent shot at an academic career. To obtain a tenured (that is, lifetime) position at a university, an assistant professor needs, in addition to excellent recommendations from colleagues, and even more importantly, a list of publications in highly ranked technical journals. Biss had such a list: the nucleus of his dissertation appeared in 2003 in the (almost) uncontested best journal in mathematics, *Annals of Mathematics*, which we've already encountered as the publisher of Wiles's proof of Fermat's Last Theorem. A second paper, which turned to the "complex" version of the results, where the earlier paper had dealt on with the "real" part, appeared in *Advances of Mathematics*, the journal founded by Gian-Carlo Rota and still edited at MIT. The papers were acclaimed by, among others, the *Mathematical Reviews* (a journal that prints short reviews of all research papers published in mathematics), which said that this was a "landmark paper" and that the main result came "as a bomb-shell." A further paper by Biss (with a coauthor, Benson Farb) appeared in 2006 in the *Inventiones Mathematicae* (founded in the 1960s by Reinhold Remmert, a professor at the University of Münster, and edited by him for decades), which surely also ranks among the best journals of math-ematics. However, such prominent publications are also sub-ject to more thorough examination than ordinary ones. And it turned out that all three of these publications had criti-cal, and probably irreparable, errors. In the case of the first paper, it was a colleague from St. Petersburg, Nikolai Mnëv, who found a critical problem in a detail and was not to be put off by the exigencies of a political campaign. Four days after the end of the campaign, Biss submitted an *Erratum* to the *Annals*, indicating an error in the paper; it was published in 2009. The *Erratum* acknowledged the error that Mnëv had found and admitted that "this affects the proof of the main theorem, Theorem 1.1, which therefore remains a conjec-ture." A defeat, which was followed by two more. Biss wrote errata for the papers in the *Advances* and the *Inventiones*, whose results may in fact be true but no valid proof has been found yet.

The fact that Biss did not submit the errata until after the end of the campaign was surely due to the efforts required to conduct the campaign, which did not leave time for the pursuit of mathematics. On the other hand, an admission of errors in his mathematics could have been awkward for a political candidate. A political opponent could well have made hay with such an admission. Unless, of course, one does not really have to worry that Republican candidates in Illinois fol-low the details of debates in the technical literature of math-ematics—and indeed, Biss won the next election handily, with 55% of the vote, and without anyone complaining about the published errata.

Caroline Lasser: Colleague

As we all know, the image of a mathematician in the general public is not terribly attractive. If we want to confirm this scientifically, we need to turn to sociologists. (This is not something we mathematicians would otherwise do.)

In May 2008, a new study by the Institute for Policy Studies in Education of the London Metropolitan University was announced to the world. Entitled "Mathematical Images and Identities: Education, Entertainment, Social Justice," it was by three British sociologists and was based in part on a poll of hundreds of students in high schools and colleges. The not very surprising result was that

> Many students and undergraduates seem to think of mathematicians as old, white, middle-class men who are obsessed with their subject, lack social skills and have no personal life outside maths.
> The student's views of maths itself included narrow and inaccurate images that are often limited to numbers and basic arithmetic.

So we have two problems: first, the image of mathematicians, and, second, the image of mathematics. These are weighty problems because they are not confined to England and because these images turn bright students, boys as well as girls, away from studying mathematics—or physics or engineering. And that is a pity, not only for the sciences but also for the long-term prospects of our society and economy. It is well known that there is already a serious shortage of scientists and engineers, and we also need many brilliant minds for mathematics, because mathematics is a key science, a foundation for much modern technology.

But it should not be surprising if a bright sixteen-year-old high-school student does not want to become like John Nash, the schizophrenic mathematician who is the subject of the film *Beautiful Mind* and who won the Nobel Prize for Economics, or like Grigori Perelman, the ingenious Russian who solved the Poincaré Conjecture but then withdrew, unshaven, to the forests outside St. Petersburg.

I have hopes that the image people have of mathematicians is becoming a little more varied and will start including images of women as well as men. It is, after all, an excellent option for a bright sixteen-year-old girl to study mathematics and contemplate a career in academia, government, or industry. The option is relatively new, the result of a hard-fought struggle in the last 200 years, and may therefore be even more interesting.

Sophie Germain (1776–1831), the daughter of a Parisian merchant, was not allowed to study mathematics, and— assuming that she would not have been taken seriously as a woman—used a male pseudonym in 1804 in order to correspond with Gauss about her researches in number theory in connection with Fermat's Last Theorem. Gauss is supposed to have made an effort to have an honorary doctorate awarded to Sophie Germain, but the efforts were made moot by her death at age fifty-five.

In 1874 Sonya Kovalevskaya submitted her doctoral thesis to the University of Göttingen, because in Berlin it was inconceivable to award a degree to a woman. She was thus the first woman in Europe with a PhD in mathematics and was later the first woman with a professorship in mathematics.

And fifty years later, Emmy Noether had to announce her lectures under David Hilbert's name and was limited to a position as Hilbert's assistant. An attempt to obtain an exemption to a regulation from 1908 that precluded women from a teaching position was not successful. The German Ministry of Education decreed in 1917, "The admission of women to teaching positions in universities continues to meet with serious objections." In response, Hilbert is reported to have said, a university department is not a bathhouse—a comment that is even more appropriate because Noether was said to have held seminars at a public pool in the summer.

Another fifty years later, in 1963, the mathematician Olga Taussky-Todd was named "Woman of the Year" by the *Los Angeles Times*. A notable sign of changing times, even if Olga Taussky remarked that this was the only honor she could have received of which none of her colleagues could be jealous—for they were all men.

Another forty-five years later, during the International Year of Mathematics, 2008, a volume of photos and interviews was released: *Frauen die forschen* (or, *Women in Research*) with portraits of twenty-five female scientists. It is a noteworthy project, introducing us to twenty-five active German researchers in a wide variety of fields including sociology, German literature, clinical medicine (fields that fit the stereotypes) but overwhelmingly natural scientists. Particularly gratifying is the inclusion of five mathematicians, which proves that mathematics has in the meantime become a viable career option for women—and, one can hope, also in turn promotes the viability of the option. Three of the five mathematicians are colleagues in Berlin. Olga Holtz is a self-assured Russian with a joint appointment at the Technical University Berlin and at UC Berkeley, a modern version of Sonya Kovalevskaya. Susanne Albers, who works in computer science and has won a Leibnitz Prize, recently came to the Humboldt University in Berlin from the University of Freiburg. And Caroline Lasser, thirty-four years old, was professor of applied mathematics at the Freie Universität Berlin; she moved on to a professorship at the Technical University of Munich recently. Munich is, in fact, the hometown for both of us, and she grew up in the same suburb where I grew up, only a few hundred meters from my house. We must have played in the same playgrounds, just twelve years apart; we attended the same elementary school on Turnerstrasse and graduated from the same high school, the Ernst Mach Gymnasium, where we had many of the same teachers. We both went to the Ludwig Maximilians Universität in Munich.

If one reviews Lasser's scientific résumé, one may or may not notice a pair of gaps in scientific output, caused by the births of her two daughters. In 2005 she was appointed group leader at the research center Matheon "Mathematics for Key Technologies" in Berlin but had to take a leave for a few months for the first baby. And again in 2008, when she was appointed to the professorship at the FU Berlin, one might find a few months' pause for the second child. In both cases she returned in a few months, with brilliant mathematical ideas concerning the computation of molecular structures. One of the earliest triumphs of quantum mechanics was the computation of the structure of atomic hydrogen (one proton and one electron); later it became possible to compute more complicated structures such as molecular hydrogen (two protons and two electrons). But as the atoms or molecules get larger, the complexity of the calculations grows incredibly fast. Lasser works with complex molecules having many different nuclei and many electrons that have some very slow motions and some very rapid oscillations. The chemists in the next building over on the FU campus are very interested in her work. To tame such large molecules computationally (and not on a supercomputer, but on a laptop!) and to understand their motions, vibrations, and excited states, one must solve incredibly complex differential equations. Even large computers grind to a halt when the molecules become too large, unless one can provide initial conditions, trial solutions, simplifications, and ideas to reduce the complexity of the computations—but always keeping in mind that one has a real molecule one is considering. This is what Lasser is working on. But since the results mean little unless one can also visualize them, she is also developing new ideas for that.

Her results are astounding, all the more since she works with more impediments than the average male mathematician.

Even if her husband, Oliver Deiser, also a mathematician, shares the load (and also the joy), the career path for a young woman with two children is not as easy as for a man. She is surely right to emphasize that also.

A cheerful young mathematician, with two children and a husband in the background, who is making her way in science and producing remarkable achievements. May I say that I truly admire that? Congratulations!

Chapter 10

What Mathematicians Can Do

The German Hair Stylists League for several years ran ads with the slogan, "What hair stylists can do, only hair stylists can do." You can probably estimate the truth value of that statement pretty well yourself. I claim, "What mathematicians can do, only mathematicians can do."

Self-Confidence and Visions

"Whoever has visions, should see a psychiatrist," the former German chancellor Helmut Schmidt used to say. Was he right? Well, maybe. But every day we are getting used to some new marvel of technology: mobile telephones, GPS navigation systems, teleconferences, interactive whiteboards, project planning at the touch of a button—all these technical systems are so complicated that no single person can design or understand them. For some of them, we assume that a good amount of mathematics went into them, but the math stays hidden behind the user interface. But isn't this a triumph of mathematics, that it is invisible in its application?

When I went to school in the 1970s, my impression was that the world of technology was in principle understandable. Not that I myself understood it all, but it seemed that one could at least in principle understand the basic ways in which, say, a TV set or a pocket calculator worked; some high school kids even understood how car engines worked. But since then, we seem to have crossed some kind of a threshold to a new level of complexity. And that was predictable, at least for visionaries. As, for example, the Briton Arthur C. Clarke (1917–2008), whose Third Law says:

> Any sufficiently advanced technology is indistinguishable from magic.

Clarke is probably best known as the author *2001: A Space Odyssey*. His three laws come from his book *Profiles of the Future: An Inquiry into the Limits of the Possible* (1962). Although he is the author of a great many novels of science fiction, Clarke is no fantasist. In 1945 he invented geostationary communications satellites. He was one of the people who really understood technology. I am also claiming him as "one of us" mathematicians: after World War II he studied physics and mathematics at King's College in London.

An opposite viewpoint to Clarke's vision is a diagnosis by the French mathematician René Thom:

> Geometry is successful magic.

A more complete version of the comment, to which Thom even adds a converse:

> The purely formal language of geometry describes effectively the reality of space. In this sense we can say that *Geometry is successful magic.*
>
> But I should immediately add a converse: Is not all magic, insofar as it is successful, geometry?

René Thom (1923–2002) is not just anyone: he is a French mathematician, an important topologist, and the creator of Catastrophe Theory, which caused a bit of a furor in the 1970s. The theory is a specialty discipline in mathematics, and it probably became a Big Thing for a while because of its name but has left hardly any traces behind. In the 1980s the attention of the media turned from Catastrophe Theory to Chaos Theory, which is responsible for any number of lovely multicolored pictures as well as the Butterfly Effect and fundamental insights into the incomputability of the real world. When the weather forecast (once again) misses the mark, the Butterfly Effect serves as a good excuse: the fluttering of a butterfly's wings in Cuba can so distort the air flows that the weather in Europe will be changed a few days later. But in practice, the weather forecasts are bad simply because there's not enough mathematics in them yet.

What's next, after Catastrophe Theory and chaos research? I claim and hope that it will be some solid mathematics: perhaps *combinatorial optimization*, a winning combination of logistics and planning, used, for example, in developing traffic networks, optimizing railway timetables, and making efficient adjustments when there are delays or breakdowns; or possibly *scientific computing*, which is used in making totally new materials or in designing drugs fitted exactly to a single individual for diagnosis or therapy.

And what encourages me in this hope? For one thing, the word "mathematics" did not appear even once in the 2006 "High-Tech Strategy" document of the German federal government, but just one year later, a pamphlet put out by the research ministry celebrated mathematics as a "Key to High Technology." It seems that politicians can learn quickly, at least in some instances.

In fact, one does not need a vision into the future to see this; this future is already here.

"Unfortunately Difficult" vs. "The Right Stuff"

Mathematics is difficult. One could, of course, try to base instruction in mathematics on the message that it's all quite simple; indeed, that approach has been tried repeatedly. But it's not true. Mathematics is not simple. Mathematics is difficult. My colleagues and I would not find doing mathematics nearly as attractive if it were easy. Sadly for the textbook writers, mathematics is and remains difficult. But we should treat that fact as a challenge and emphasize it almost like a marketing slogan. My inspiration for this is a marketing campaign in the early 1990s by the clothing brand René Lezard, which used the slogan:

> You don't need a perfectly fitting suit to become
> Finance Minister. But the investment would be useful.
> René Lezard. Unfortunately expensive.

Mathematics. Unfortunately difficult. We have to be proud of that "unfortunately" and say it defiantly. But we must also remember that "difficult" is relative, even if it is true for everyone.

I have a poster in my office with a picture of Einstein and a quote:

> Don't worry about your difficulties in mathematics. I
> can assure you, mine are greater.

The quote is really too cute. Could Einstein really have said or written that? In fact, he did. It was in a letter to Barbara Wilson, a high-school student, that he wrote on January 7, 1943. We can learn from that letter not only that math has been a difficult subject in high schools for a long time but also that Einstein made time to answer letters from high-school students.

If Einstein, the physicist with the fright-wig hairdo, does not seem like a suitable role model, we can turn to the Hungarian-American John von Neumann:

> If people do not believe that mathematics is simple, it is only because they do not realize how complicated life is.

Although that quote also seems too good to be true, it has a good pedigree: it comes from a discussion after von Neumann's keynote address to the first meeting of the Association for Computing Machinery in 1947. And von Neumann is clearly a superlative role model. Every photo of him shows him with a perfectly fitting suit.

OK, then: unfortunately difficult. I'll expand that to "unfortunately difficult, colon." The "difficult" has been confirmed by Einstein and von Neumann, which is reassuring. The "unfortunately" we've purchased from René Lezard's ad agency and have agreed to say it with full confidence. But then we should not follow it with an exclamation point. We should not say "unfortunately difficult, exclamation point," as if that were it, nothing more to say, but rather, "unfortunately difficult, colon," because after the colon it really starts to get interesting. That "unfortunately difficult" is exciting for a mathematician. Mathematics offers problems, challenges. Are there enough young people who find that exciting? Or do we have to make special efforts to entice young people into mathematics, for example, in schools? That question was put to mathematician Jean-Pierre Serre in an interview in 1986. Serre is an outstanding French mathematician who was awarded a Fields Medal in 1954 and the first Abel Prize from the Norwegian Academy of Sciences in 2003. Serre gave the following answer:

> I have a theory about this, which says that we should initially advise people *not* to do mathematics. One

does not need all that many mathematicians. However, if they still insist on doing mathematics, then one should indeed encourage them and help them.

In at least part of his response, Serre was in error, or else the situation has changed dramatically since 1986: there is indeed a need for a large number of mathematicians—and not just for old white males who look like Einstein.

Mathematics is difficult. Some of the problems may be the most difficult intellectual problems humans can face. As a battle cry for dealing with them, I will quote from the famous lecture that David Hilbert presented at the 1900 International Congress of Mathematicians. In that talk he formulated the mission of our field thus:

> We hear in us the constant call: Here is a problem, seek the solution. You can find it through pure thought alone, for in mathematics there is no "ignorabimus."

And we hope that the most talented young people, of all backgrounds, will rise to that challenge. Mathematics is a field for heroes. We have to recruit young people who will dare to work in mathematics.

Tom Wolfe describes the essence of heroism in his documentary novel *The Right Stuff* about the early days of the U.S. space program (the book was published in 1979; the film came out in 1983). "The right stuff" refers to the combination of recklessness, chutzpah, willingness to take risks, heroism, and arrogance that each of the pilots had to have—and had to *demonstrate*—if he wanted to succeed. In the second chapter of his book, with the heading "The Right Stuff," the mathematical metaphors jump out at us. Wolfe describes the heroism of his pilots with many mathematical images:

■ The training of pilots is like a *pyramid* on which one climbs steadily upward.

- The flight deck on which one has to land is, in the training film, "a big piece of gray *geometry*, perilous, to be sure, but an amazing *abstract* shape."
- That there are no unpredictable accidents is presented as a *theorem* and *demonstrated*, and *corollaries* were derived. One of the theorems was that there were "no accidents and no fatal flaws in the machines, there are only pilots with the wrong stuff."
- But that is never stated explicitly, only in *code* or by *example.*
- And in the end comes *statistics*: "Once the theorem and the corollary were understood, the Navy's statistics about one in every four Navy aviators dying meant nothing. The figures were *averages*, and the averages applied to those with the average stuff."

We don't heed the heroism of fighter pilots and astronauts in mathematics. Our heroes can be peaceable and remain on the ground. "Unfortunately difficult, colon" is an invitation to young people who would like to set their ambition and their "certain something" to the challenges of an exciting and difficult discipline. And they can be proud when they succeed.

Record Races

Chasing after records has never been a major preoccupation of mathematicians. However, there are one or two such competitions that have been going on for millennia. There is, for example, the competition for computing the largest number of digits in the decimal expansion of π. The first to reach more than ten correct digits was Al Kashi in 1429; by 1706 Machin had already achieved a hundred digits; and in 1949 Smith and Wrench produced over a thousand digits—using a desktop mechanical calculator.

Why do that? Aside from the fact that newspapers (and university press offices) always like a story about a new record being set, the effort is a sport, a challenge, and a competition. A few such records are even recognized with the award of a prize. An additional reason, and a much more serious and sensible one, is that the record-breaking computation (or whatever) demonstrates the progress of the theory and shows what is currently practicable. Someone who wants to compute a million digits of π will find it impossible using only formulas known in the eighteenth century. The new formulas that make it possible to compute a million digits did not, of course, fall from the sky, but are the result of serious research in number theory.

The current record for computing the decimal expansion of π was set on October 17, 2011, by a pair of mathematicians, Alexander J. Yee and Shigeru Kondo, who computed it out to 10 trillion digits. Like the previous record-setting computation, by Fabrice Bellard (December 31, 2009, 2.7 trillion digits), the computation used a desktop computer, not a supercomputer, which had been used for all of the earlier record-setting computations after supercomputers became available. Yee and Kondo report that it took over a year to accomplish the feat, with occasional hard-drive failures slowing the computations down. Both Yee and Kondo as well as Bellard were able to make use of a particularly elegant formula discovered by two Russian-American brothers, David and Gregory Chudnovsky, in 1987—and used by them in 1989 to set a new record for digits of pi that held for eight years.

Even if (as I suspect) you are not tempted to participate in such programming challenges as these, you can nonetheless participate in another project that seeks to set records. In this case, it's the search for especially large prime numbers. The Electronic Frontier Foundation has set up some awards for the project, including one for $150,000 for the discovery of a prime number with more than a hundred million digits.

The project is called GIMPS (Great Internet Mersenne Prime Search, www.mersenne.org) and uses tens of thousands of home computers to test candidate prime numbers. The numbers are all of the form $2^p - 1$, where p is itself a prime, so these primes are "Mersenne primes," such as $M_2 = 2^2 - 1 = 3$, $M_3 = 2^3 - 1 = 7$, $M_5 = 2^5 - 1 = 31$, or $M_7 = 2^7 - 1 = 127$. Numbers of this form can clearly be prime numbers, but they need not be (for example, $M_{11} = 2^{11} - 1 = 2047 = 23 \times 89$ is not a prime). Why limit the search to Mersenne primes? For one thing, they are easy to write down, and, for another, they are easier to test for primality than "general" numbers with several million digits. And maybe we can hope for some divine inspiration. Mersenne was, after all, a monk.

If you want to participate in GIMPS, you can download the software to test numbers for primality from the GIMPS website; you will then be assigned some candidate numbers that you can feed to your computer to keep it busy and out of mischief during idle hours. If you're lucky, one of your candidates will be a new Mersenne prime. The largest currently known prime number is $M_{43,112,609} = 2^{43,112,609} - 1$, a number with nearly 13 million digits; it was found by a computer in the GIMPS project on 23 August 2008. And since it was the first prime number discovered with more than ten million digits, the Electronic Frontier Foundation paid out an award of $100,000. (The tragic near-hero of the story is the prime-number enthusiast Hans-Michael Elvenich from Cologne, Germany, who runs the website www.primzahlen.de. He found another prime number of almost the same size, also using the GIMPS software, but a few days later, and got little recognition and no prize.)

Time magazine named the record-breaking number $M_{43,112,609}$ as one of the 50 Best Inventions of 2008—to be precise, it was number 29 in the list, an appropriate prime number. (Number 1 was a retail DNA test that would be available to individuals in drugstores.) *Time* describes the search for ever-larger Mersenne primes as the unofficial "national" sport of mathematicians. Yes, it is correct, but isn't it rather an "international" sport?

You Know More Math Than You Think

Who among us uses math in our everyday lives? In general, we don't do much computing—at least, not in the sense of working with numbers to get some result. Of course, we count from time to time (but often only up to four, like the bees), and we sometimes add up prices in the supermarket or in the restaurant (so as not to be too surprised by the final bill). Computing percentages is useful in many situations, and occasionally we deal with probabilities. The reading of time-tables is a clearly mathematical activity. It was praised by the poet Hans Magnus Enzensberger: "Read no odes, my son, read timetables, for they are more precise," he writes in his 1957 poem "Into the Storybook for the Upper Grades." We read timetables critically, taking note of unnecessary detours, seeking the best, *optimized*, path and connections that begins at the desired place and time: this is mathematics, *combinatorial optimization* in everyday use.

A tourist in Frankfurt sees a weather forecast for tomorrow that is divided into two parts: sunny weather is predicted for the north and west of Germany, clouds and rain for the south and east, with a border that runs diagonally across Germany. Frankfurt and Leipzig are just on the dividing line. What does this mean for our friend the tourist, and what should he believe about this forecast anyway? A larger weather map of Europe shows a huge low over the Balkans and a stable and undeserved high over the British Isles; the border between them is said to be moving hardly at all—however, one can see ("at a glance") that the border is totally unstable and can, and probably will, move one way or another unpredictably. Congratulations to our friend: he has just done some sophisticated math. *Stability* is a complicated mathematical concept, and computing the stability or instability of a system is difficult. Someone who sees the simple forecast and mistrusts the apparent precision based on past experience, and then sees how the weather might actually behave (on the map) is using mathematics, *scientific computation.*

Automobile bodies, even for racing cars, are rarely tested in wind tunnels any more. Computerized virtual wind tunnels have become an important design tool and an alternative to expensive tests with real models in real wind tunnels. The computations, however, are exceedingly difficult: one has to be able to compute vortices and turbulence correctly to understand and minimize them. When this works, one can see immediately if a car is streamlined, and if the dealer shares the video with you, you can ignore the sales pitch and promo text in the brochures; you can instead look at the details of the streamlines shown to you by the expensive high-performance parallel computer at the car company's R&D center and readily judge the aerodynamics of the car. This too is mathematics in everyday life: *numerical solutions of differential equations.*

All these are good reasons for mathematical confidence: we have a good feel for complicated concepts such as *optimal, stable,* and *streamlined.* When we deal with concepts such as *probable* and *likely,* when we think of a value as *average* or *extreme,* we generally do that in the back or our minds (or, in some cases, even in the brainstem), but we are indeed doing mathematics: rapidly, even intuitively, and largely error-free.

"Mathematics Is ..."

I think that most statements about "mathematics" are either false or devoid of content, just like statements about "Americans" or "love."

Like the marketing slogan "I am two oil tanks" from the 1970s, the heading "Mathematics" covers a wide variety of themes: mathematics as a subject in schools, mathematics as a tool for everyday use, mathematics as a cultural achievement, mathematics as an indispensable foundation for science and engineering, and mathematics as a subject for research. That's already five completely different kinds of mathematics.

And if it's claimed that "mathematics is difficult," that's certainly true for the science but irrelevant for the cultural achievement. And if I claim that "mathematics is easy," then I'm referring to the school subject and the everyday activity, but not the research excursions. In the same vein, it is also true that "You know more math than you think," which was the slogan of the Year of Mathematics 2008.

Because mathematics is so multifaceted, and because so many different things are mathematics, it is difficult to make general, true, and meaningful statements that start, "Mathematics is …." One can, of course, say "mathematics is important," and that's unassailable, as is "mathematics is (unfortunately) difficult." So maybe we should start collecting statements about the theme "Mathematics is …," as a cartoon series did with "Love is …." Woody Allen claims that "Love is dirty if it's done right." Maybe it's also true that "Mathematics is dirty if it's done right." But I don't want to be quoted on that.

Read More

I am glad about (nearly) all the stories about mathematics that I had the opportunity to experience myself. On the other hand, I thank everyone who told me stories that I could retell in this book. Many further stories and details I found through research—often on the Internet, and especially with support from Wikipedia (both the German and English versions). Some other essential and critical resources to which I could track the foundations of the stories or that I came across and plowed into for this book are listed below.

In this list I've marked a dozen books with bullets ▶; these are books in which I found the inspiration to retell the stories that make up this book. The bullets ▶ thus imply a particularly strong recommendation to "Read this!"

Preface

Richard Courant and Herbert Robbins, *What Is Mathematics? An Elementary Approach to Ideas and Methods,* Oxford University Press, 1941 (2nd Edition, 1996).

On the Number Line
3—Can Bees Count?

Hans J. Gross, Mario Pahl, Aung Si, Hong Zhu, Jürgen Tautz, and
Shaowu Zhang, "Number-based visual generalization in the
honeybee," *PLoS One*, (1) 4, (2009), e4263, www.plosone.org.
Marie Dacke and Mandyam V. Srivisan "Evidence for counting in
insects," *Animal Cognition* 11 (2008), pp. 683–689.
Oliver Deiser, *Einführung in die Mengenlehre—Die Mengenlehre
Georg Cantors und ihre Axiomatisierung durch Ernst Zermelo*,
Springer, Heidelberg, 2. Auflage 2004. A nice reference in English
is Keith Devlin, *The Joy of Sets*, Springer, New York, 2nd ed., 1993.

10—And the Name of the Rose

► Hans Wussing, *6000 Years of Mathematics: A Cultural-
Historical Journey*. Two volumes, Springer, Heidelberg,
2008/2009.
Lothar Gerritzen, ed., Manfred Hauenschild, Paul Kimmeskamp
and Jürgen Voigt, Zwanzigeins. *Für die unverdrehte
Zahlensprechweise*, Universitätsverlag Dr. N. Brockmeyer, 2008.
Octomatics, www.octomatics.org.

13—Bad Luck?

Edgar Wunder, "Die Folgen von 'Freitag, dem 13.' auf das
Unfallgeschehen in Deutschland," *Zeitschrift für Anomalistik* 3
(2003), pp. 47–55; www.skeptizismus.de/freitag.pdf.

42—The Answer to Everything?

► Douglas Adams, *The Hitchhiker's Guide to the Galaxy/The
Restaurant at the End of the Universe*, Del Rey, New York, 2005.

91—The Numbers on the Bone

Folkmar Bornemann, "PRIMES is in P: A breakthrough for 'everyone,'" *Notices of the American. Mathematics Society* 5, no. 50
(May 2003), pp. 545–552, http://www.ams.org/notices/200305/
fea-bornemann.pdf.

π—As Beautiful as the Mona Lisa?

Daniel Tammet, *Born on a Blue Day: A Memoir of Asperger's and an Extraordinary Mind*, Hodder and Stoughton, London, 2006.

Lennart Berggren, Jonathan Borwein, and Peter Borwein, *Pi: A Source Book*, Springer, New York, 1997 (3rd Edition, 2004).

$\sqrt{-1}$ —Victim of a Character Assassination

Heinz-Dieter Ebbinghaus et al., eds., *Numbers*, Graduate Texts in Mathematics, Springer, New York, 1990.

The Never-Ending Story of Prime Numbers

Neil J. A. Sloane, *On-Line Encyclopedia of Integer Sequences*, www.research.att.com/~njas/sequences/.

Euclid Is still Right

Benno Artmann, *Euclid: The Creation of Mathematics*, Springer, Heidelberg, 1999.

How Many Prime Numbers Are There?

▶ Marcus du Sautoy, *The Music of the Primes*, Harper-Collins, New York, 2003.

Paulo Ribenboim, *The New Book of Prime Number Records*, Springer, New York, 1988 (3rd Edition, 1996).

The "Mozart of Mathematics" Makes Use of an Error

Barry Cipra, "New insights into prime numbers," in *What's Happening in the Mathematical Sciences 1995–1996*, Vol. 6, American Mathematical Society, Providence, RI, 2006, pp. 52–65.

Another Search for Errors

Barry Cipra, "Divide and conquer," in *What's Happening in the Mathematical Sciences 1995–1996*, Vol. 3, American Mathematical Society, Providence, RI, 1996, pp. 38–47.

A Well-Insured Million

Apostolos Doxiadis, *Uncle Petros and Goldbach's Conjecture: A Novel of Mathematical Obsession*, Bloomsbury, New York, 2000.

The Mathematical Perspective

Random Numbers

Bernd Beber and Alexandra Scacco, "The devil is in the digits," *Washington Post*, 20 June 2009.
Boudewijn F. Roukema, "Benford's law anomalies in the 2009 Iranian presidential elections," Preprint, version 3, 26 June 2009, 25 pp., arxiv.org/abs/0906.2789.

Everything Far Above Average

Garrison Keillor, *Lake Wobegon Days*, Viking, New York, 1986.

Caution: Equations

The Huntington Affair

Ivo K. Feierabend and Rosalind L. Feierabend, "Aggressive behaviours within politics 1948–1962: A cross-national study," *Journal of Conflict Resolution* 10, no. 3 (September 1966), pp. 249–271.
Samuel P. Huntington, *Political Order in Changing Societies*, Yale University Press, New Haven, 1968.
Neil Koblitz, "Mathematics as propaganda," in L.A. Steen, ed., *Mathematics Tomorrow*, Springer, New York, 1981, pp. 111–120.

Serge Lang, "Academia, journalism, and politics: A case study," in *Challenges*, Springer, New York, 1998, pp. 1–222.

Pythagoras Lives

Douglas Adams, *Dirk Gently's Holistic Detective Agency*, William Heinemann, London, 1987.

Equations as Art

► Craig Damrauer, *New Math: Equations for Living*, Andrews McMeel Publishing, Kansas City, MO, 1987.

The Small Puzzles

Sudokus

Gordon Royle, *Minimum Sudoku*, units.maths.uwa.edu. au/~gordon/sudokuid.php.
DFG-Research Center Matheon, Sudoku, www.matheon.de/schools/sudoku/sudoku.asp.

3x + 1

Jeffrey C. Lagarias, "The $3x + 1$ problem and its generalizations," *American Mathematical Monthly* 92 (1985), pp. 3–23.

The Perfect Monster

David G. Kendall, "The scale of perfection," *Journal of Applied Probability* 19A (1982), pp. 125–138.

The Great Puzzles

► Simon Singh, *Fermat's Enigma*, Walker, New York, 1997.
Christoph Pöppe, "Der Beweis der Catalan'schen Vermutung," *Omega (Spektrum der Wissenschaft Spezial)* 4 (2003), pp. 64–67.
Clay Mathematics Institute, *Millennium Problems*, www.claymath. org/millennium/.

Where Mathematics Is Created

At the Café

R. Daniel Mauldin, *The Scottish Book: Mathematics from the Scottish Café*, Birkhäuser, Boston, 1981.

Ariel Rubinstein, *A Worldwide Guide for Coffee: Places Where You Can Not Only Work but Also Think!*, arielrubinstein.tau.ac.il/univ-coffee.html.

► Karl Sigmund, John Dawson, and Kurt Mühlberger, *Kurt Gödel: Das Album—The Album* (German and English Edition), Vieweg, Wiesbaden, 2006.

In the Computer

Jonathan M. Borwein, Peter B. Borwein, and Karl Dilcher, "Pi, Euler numbers, and asymptotic expansions," *American Mathematical Monthly* 96 (1989), pp. 681–687.

In Bed

Clemens Schaefer, ed., *Briefwechsel zwischen Carl Friedrich Gauβ und Christian Ludwig Gerling*, O. Elsner, Berlin, 1927.

G. H. Hardy, *Ramanujan. Twelve Lectures on Subjects Suggested by His Life and Work*, Cambridge University Press, Cambridge, 1940.

Robert Kanigel, *The Man Who Knew Infinity: A Life of the Genius Ramanujan*, Scribner, New York, 1991.

In Captivity

Karl Sigmund, Kühler, "Abschied von Europa—Wien 1938 und der Exodus der Mathematik," Österreichische Mathematische Gesellschaft, Vienna, 2001. www.oemg.ac.at/Tagungen/2001/KuehlerAbschied.pdf.

Peter Weibel and Friedrich Stadler, *Vertreibung der Vernunft: The Cultural Exodus from Austria*, Locker Verlag, Vienna, 1993.

Anna Maria Sigmund, "Peter Michor and Karl Sigmund, Leray in Edelbach," *Internat. Math. Nachrichten* no. 199 (2005), pp. 1–16.

André Weil, *The Apprenticeship of a Mathematician*, Birkhauser, Boston, 1992.

In an Attic Room in Princeton

Simon Singh and John Lynch, *Fermat's Last Theorem*, documentary, BBC, 1996.
Simon Singh, *Fermat's Enigma, The Epic Quest to Solve the World's Greatest Mathematical Problem*, Walker, New York, 1997.

On a Beach

Martin Grötschel, ed., *The Sharpest Cut: The Impact of Manfred Padberg and His Work*, SIAM, Philadelphia, PA, 2004.
Stephen Smale, "The story of the higher dimensional Poincaré Conjecture (what actually happened on the beaches of Rio)," *Mathematical Intelligencer* 2, no.12 (1990), pp. 44–51.

Knowledge in the ArXiv

ArXiv Mathematics, arxiv.org/new/math.html.
George G. Szpiro, *Poincaré's Prize: The Hundred Years' Quest to Solve one of Math's Greatest Puzzles*, Dutton, 2007.

Research in the Internet?

Timothy Gowers, *Mathematics: A Very Short Introduction*, Oxford University Press, Oxford, 2002.
Timothy Gowers, "Is massively collaborative mathematics possible?" on Gowers's Weblog: gowers.wordpress.com, 27 January 2009.

The Book of Proofs

Martin Aigner and Günter M. Ziegler, *Proofs from THE BOOK*, Springer, Heidelberg, 2010 (4th Edition).

About Computer Proofs

Thomas C. Hales, Georges Gonthier, John Harrison, and Freek Wiedijk, "Formal proof," *Notices of the American Mathematical Society* 55, no. 11 (2008), pp. 1370–1414, www.ams.org/notices/200811/.
Doron Zeilberger, *Dr. Z's Opinions*, www.math.rutgers. edu/~zeilberg/OPINIONS.html.

Concerning Precision

Jonathan Borwein and Keith Devlin, *The Computer as Crucible: An Introduction to Experimental Mathematics*, A K Peters, Natick, MA, 2009.
Ed Sandifer, *How Euler Did It*, Mathematical Association of America, Washington, DC, 2007.

Concerning Surprises

Felix Mühlhölzer, "Wittgenstein and surprises in mathematics," in R. Haller, and K. Puhl, eds., *Wittgenstein und die Zukunft der Philosophie. Eine Neubewertung nach 50 Jahren*, Öbvandhpt, Vienna, 2002, pp. 306–315.
▶ G. H. Hardy, *A Mathematician's Apology*, Cambridge University Press, Cambridge, 1940.

Three Legends

Eric T. Bell, *Men of Mathematics*, Simon and Schuster, New York (1937).

Mathematician vs. Mathematician

Jacques Hannak, *Emanuel Lasker. Biographie eines Schachweltmeisters*, Siegfried Engelhardt Verlag, Berlin, 1952 (3rd Edition, 1970).

Was It Kovalevskaya's Fault?

Sonja Kowalewski, *Jugenderinnerungen*, S. Fischer, Berlin, 1897 (Neuausgabe 1983 mit Nachwort von Peter Härtling). English version: *Sonya Kovalevsky: Her Recollections of Childhood (Classic Reprint)*, Forgotten Books, http://www.forgottenbooks.org/.
Lars Gårding and Lars Hörmander, "Why is there no Nobel Prize in mathematics," *Mathematical Intelligencer* 7, no. 3 (1985), pp. 73–74.

Ulf Person, "Interview with Arild Stubhaug," *Notices of the American Mathematical Society* 52, no. 9 (October 2005), pp. 1046–1051, www.ams.org/notices/200509/comm-persson.pdf.

The Disappearance of Alexander Grothendieck

▶ Winfried Scharlau, Wer ist Alexander Grothendieck? Anarchie, Mathematik, Spiritualität. Eine Biographie. Teil 1: Anarchie, Selbstverlag, 2008, winfried.scharlau@web.de.
Winfried Scharlau, "Wer ist Alexander Grothendieck?" Oberwolfach Lecture 2006, Mathematisches Forschungsinstitut Oberwolfach 2007.

What Kinds of People Are These?

Paul Erdős: Traveler

▶ George Csicsery, *N Is a Number. A Portrait of Paul Erdős*, Documentary film, 57 min., VHS 1993/DVD 2005, Springer.
Bruce Schechter, *My Brain Is Open: The Mathematical Journeys of Paul Erdős*, Simon and Schuster, New York, 1998.

Gian-Carlo Rota: Provocateur

Mark Kac, Gian-Carlo Rota, and Jacob T. Schwartz, *Discrete Thoughts*, Birkhäuser, Boston, 1986.
▶ Gian-Carlo Rota, *Indiscrete Thoughts*, Birkhäuser, Boston, 1997.
Ester Gasperoni Rota, "Gianco, my brother," in E. Damiano et al., eds., *From Combinatorics to Philosophy: The Legacy of G.-C. Rota*, Springer, New York, 2009.

Persi Diaconis: Magician

▶ Donald J. Albers, ed., *Mathematical People*, A K Peters, Wellesley, MA, 2008 (2nd Edition).

Caroline Lasser: Colleague

► Jeanne Rubner, ed., *Frauen die forschen: 25 Portraits von Bettina Flitner*, Collection Rolf Heyne, Munich, 2008.

What Mathematicians Can Do

"Unfortunately Difficult" vs. "The Right Stuff"

Tom Wolfe, *The Right Stuff*, Farrar, Straus and Giroux, New York, 1979.

Self-Confidence and Visions

Arthur C. Clarke, *Profiles of the Future: An Inquiry into the Limits of the Possible*, Henry Holt and Co., New York, 1984.

Record Races

George Woltman, Scott Kurowski, et al., *Great Internet Mersenne Prime Search (GIMPS)*, www.mersenne.org.

Günter M. Ziegler, "The great prime number record races," *Notices of the American Mathematical Society* 4, no. 51 (2004), pp. 414–416, www.ams.org/notices/200404/comm-ziegler.pdf.

My Sporting Successes

Harold P. Boas, "The football player and the infinite series," *Notices of the American Mathematical Society* 44, no. 11 (December 1997), pp. 1430–1435. www.ams.org/notices/199711/boas.pdf.

Index